TM 9-803

Willys Overland Model MB and Ford Model GPW 1/4 Ton 4x4 Technical Manual

Department of Defense

Pentagon
Publishing

WAR DEPARTMENT
Washington 25, D. C, 22 February 1944

TM 9-803, ¼-ton 4 x 4 Truck (Willys-Overland Model MB and Ford Model GPW), is published for the information and guidance of all concerned.

[A. G. 300.7 (17 November 43)]

BY ORDER OF THE SECRETARY OF WAR:

G. C. MARSHALL,
Chief of Staff.

OFFICIAL:

J. A. ULIO,
Major General,
The Adjutant General.

DISTRIBUTION: C & H (1).

(For explanation of symbols, see FM 21-6)

¼-TON 4 x 4 TRUCK (WILLYS-OVERLAND MODEL MB and FORD MODEL GPW)

CONTENTS

PART ONE—OPERATING INSTRUCTIONS

Section I

INTRODUCTION

1. SCOPE.

a. This technical manual* is published for the information and guidance of the using arm personnel charged with the operation and maintenance of this materiel.

b. In addition to a description of the ¼-ton 4 x 4 Truck (Willys-Overland model MB and Ford GPW), this manual contains technical information required for the identification, use, and care of the materiel. The manual is divided into two parts. Part One, sections I through VII, contains vehicle operating instructions. Part Two, sections VIII through XXXII, contains vehicle maintenance instructions to using arm personnel charged with the responsibility of doing maintenance work within their jurisdiction, including radio suppression and shipment and temporary storage information.

c. In all cases where the nature of the repair, modification, or adjustment is beyond the scope of facilities of the unit, the responsible ordnance service should be informed so that trained personnel with suitable tools and equipment may be provided, or proper instructions issued.

d. This manual includes operating and organizational maintenance instructions from the following Quartermaster Corps 10-series technical manuals. Together with TM 9-1803A and TM 9-1803B, this manual supersedes them:

(1) TM 10-1103, 20 August 1941.

(2) TM 10-1207, 20 August 1941.

(3) TM 10-1349, 3 January 1942.

(4) TM 10-1513, Change 1, 15 January 1943.

*To provide operating instructions with the materiel, this technical manual has been published in advance of complete technical review. Any errors or omissions will be corrected by changes or, if extensive, by an early revision.

RA PD 305251

Figure 1—1/4-Ton 4 x 4 Truck—Left Front

INTRODUCTION

RA PD 305250

Figure 2—1/4-Ton 4 x 4 Truck—Right Rear

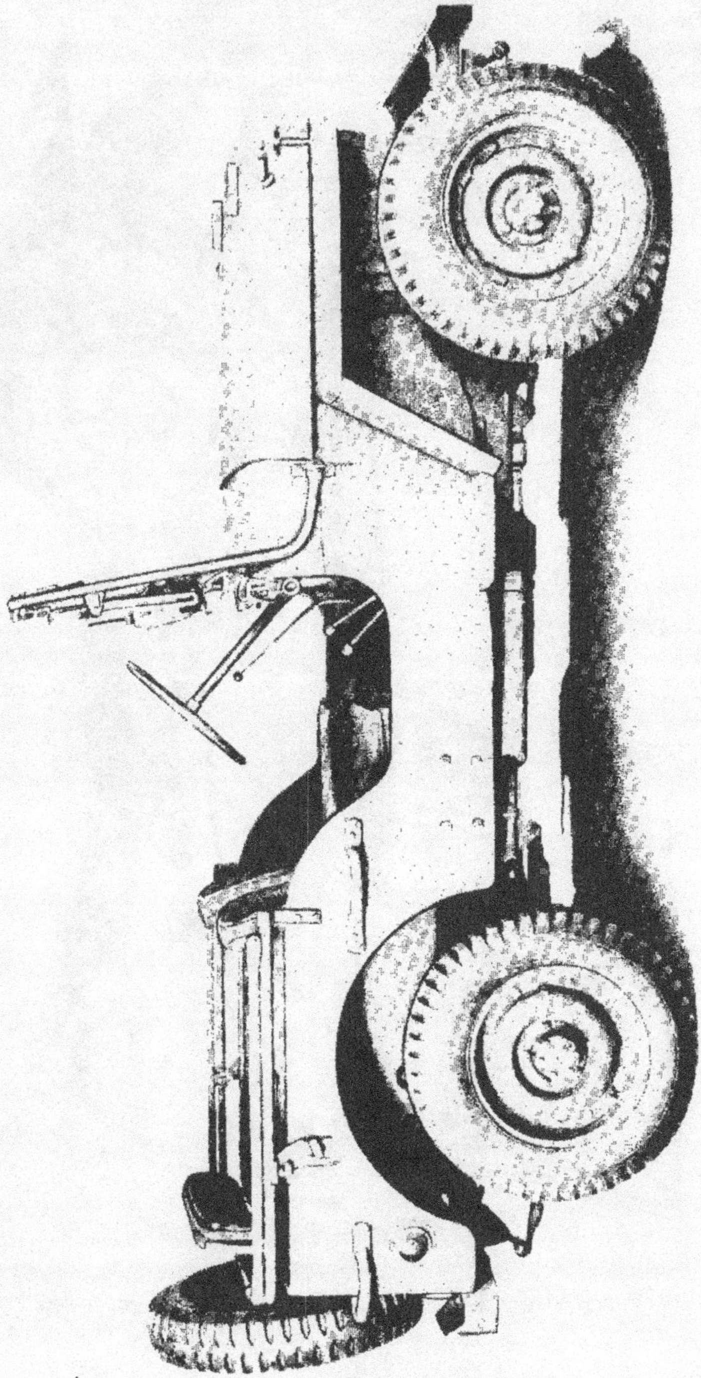

RA PD 305163

Figure 3—1/4-Ton 4 x 4 Truck—Right Side

INTRODUCTION

RA PD 305164

Figure 4—1/4-Ton 4 x 4 Truck—Right Front

¼-TON 4 x 4 TRUCK (WILLYS-OVERLAND MODEL MB and FORD MODEL GPW)

Section II

DESCRIPTION AND TABULATED DATA

2. DESCRIPTION.

a. Type. This vehicle is a general purpose, personnel, or cargo carrier especially adaptable for reconnaisance or command, and designated as 1/4-ton 4x4 Truck. It is a four-wheel vehicle with four-wheel drive. The engine is a 4-cylinder gasoline unit located in the conventional place, under the hood at the front of the vehicle. A conventional three-speed transmission equipped with a transfer case provides additional speeds for traversing difficult terrain. The body is of the open type with an open driver's compartment. The folding top can be removed and stowed; and, the windshield tilted forward on top of the hood, or opened upward and outward. A spare wheel equipped with a tire is mounted on the rear of the body, and a pintle hook is provided to haul trailed loads. Specifications of the vehicle are given under "Data" (par. 3). General physical characteristics are shown in figures 1 through 4.

h. Identification. The manufacturer's chassis serial number is stamped on a plate inside the left frame side member at the front end, and on the name plate (fig. 6). The engine serial number is stamped on the right side of the cylinder block, front upper corner. The U.S.A. registration number is painted on both sides of the hood.

3. DATA,

a. Vehicle Specifications.

Wheelbase	80 in.
Length, over-all	1321/4 in.
Width, over-all	62 in.
Height, over-all—top up	693/4 in.
—top down	52 in.
Wheel size	combat 16 x 4.50 E
Tire size	16 x 6.00 in.
Tire pressure (front and rear)	35 lb
Tire type	mud and snow
Tire plies	6
Tread (center-to-center)—front	49 in.
—rear	49 in.
Crew, operating	2
Passenger capacity including crew	5

Weights:	
Road, including gas and water	2,453 lb
Gross (loaded)	3,253 lb
Shipping (less water and fuel)	2,337 lb
Boxed gross	3,062 lb
Maximum pay load	800 lb
Maximum trailed load	1,000 lb
Ground clearance	8 3/4 in.
Pintle height (loaded)	21 in.
Kind and grade of fuel (octane rating) Gasoline (68 mm)	
Approach angle	45 deg
Departure angle	35 deg
Shipping dimensions—cubic feet 331	
—square feet 57	

b. **Performance.**

Maximum allowable speeds (mph) with transfer case in "HIGH" range:	
High gear (3rd)	65
Intermediate gear (2nd)	41
Low gear (1st)	24
Reverse gear	18
Maximum allowable speeds (mph) with transfer case in "LOW" range:	
High gear (3rd)	33
Intermediate gear (2nd)	21
Low gear (1st)	12
Reverse gear	9
Maximum grade ability	60 pct
Minimum turning radius—right	17 1/2 ft
—left	171/2 ft
Maximum fording depth.	21 in.
Towing facilities—front	none
—rear	pintle hook
Maximum draw-bar pull	1,930 lb
Engine idle speed	600 rpm
Miles per gallon—(high gear—high range) average conditions	20
Cruising range—(miles) average conditions	20

c. **Capacities.**

Engine crankcase capacity—dry 5 qt	
—refill 4 qt	
Transmission capacity	% qt
Transfer case capacity	1 l/l qt

¼-TON 4 x 4 TRUCK (WILLYS-OVERLAND MODEL MB and FORD MODEL GPW)

Front axle capacity (differential)	1¼ qt
Rear axle capacity (differential)	1¼ qt
Front axle steering knuckle universal joint	¼ qt
Steering gear housing	¼ qt
Air cleaner (oil bath)	⅝ qt
Fuel tank capacity	15 gal
Cooling system capacity	11 qt
Brake system (hydraulic brake fluid)	¼ qt
Shock absorbers--front	5 oz
—rear	5¾ oz

d. Communications.

(1) **Radio Outlet Box.** A radio outlet box is provided on the later vehicles to use the vehicle battery (6-volt current supply). This outlet is located against the body side panel at the right front seat.

(2) **Auxiliary Generator.** A 12-volt, 55-ampere auxiliary generator is furnished on some vehicles. The generator is driven by a V-belt from a power take-off unit on the rear of the transfer case. Instructions for operation and care accompany those vehicles.

Section III

DRIVING CONTROLS AND OPERATION

Paragraph

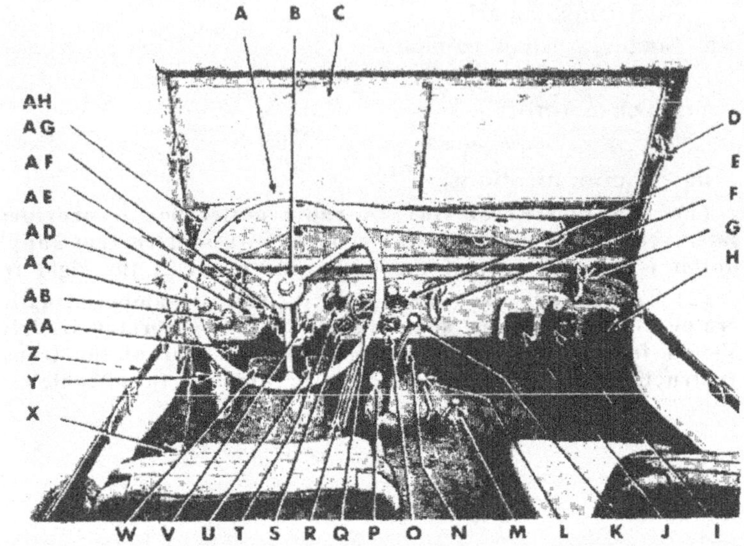

A STEERING WHEEL
B HORN BUTTON
C WINDSHIELD WIPERS
D WINDSHIELD ADJUSTING ARMS
E AMMETER
F HAND BRAKE
G WINDSHIELD CLAMPS
H CAUTION PLATE
I NAME PLATE
J SHIFT PLATE
K TRANSMISSION GEAR SHIFT LEVER
L TRANSFER CASE SHIFT LEVER—FRONT AXLE DRIVE
M TRANSFER CASE SHIFT LEVER—AUXILIARY RANGE
N STARTING SWITCH
O TEMPERATURE GAGE
P ACCELERATOR FOOT REST
Q SPEEDOMETER
R ACCELERATOR (FOOT THROTTLE)
S OIL PRESSURE GAGE
T FUEL GAGE
U BRAKE PEDAL
V INSTRUMENT PANEL LIGHT SWITCH
W CLUTCH PEDAL
X FUEL TANK
Y FIRE EXTINGUISHER
Z SAFETY STRAP
AA HEADLIGHT FOOT SWITCH (BEAM CONTROL)
AB BLACKOUT LIGHT SWITCH
AC BLACKOUT DRIVING LIGHT SWITCH
AD REAR VISION MIRROR
AE CHOKE CONTROL
AF IGNITION SWITCH
AG HAND THROTTLE
AH RIFLE HOLDER

RA PD 334753

Figure 5—Instruments and Controls

4. INSTRUMENTS AND CONTROLS,

a. **Instruments.**

(1) Ammeter (fig. 5). The ammeter on the instrument panel indicates the rate of current flow when the generator is charging the battery, and also indicates the amount of current being consumed when the engine is idle.

(2) Fuel Gage (fig. 5). The fuel gage on the instrument panel

¼-TON 4 x 4 TRUCK (WILLYS-OVERLAND MODEL MB and FORD MODEL GPW)

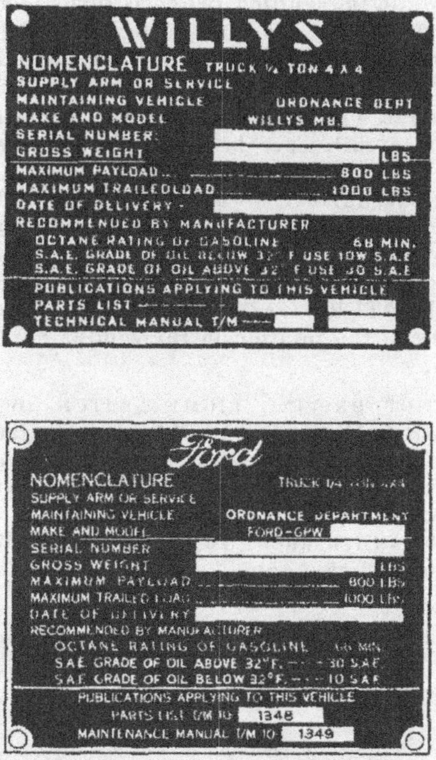

RA PD 330838

Figure 6—Name Plate

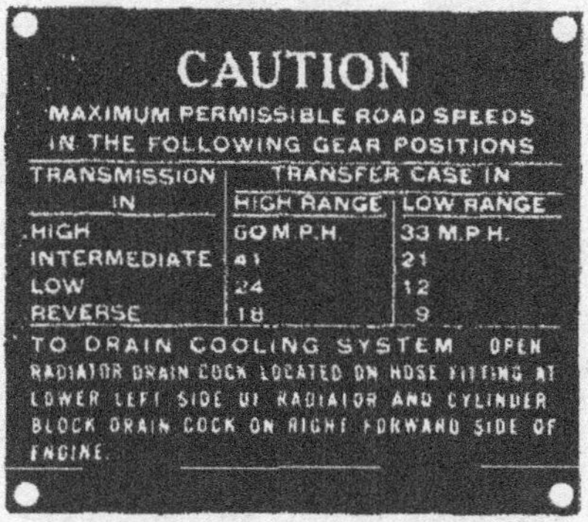

RA PD 305162

Figure 7—Caution Plate

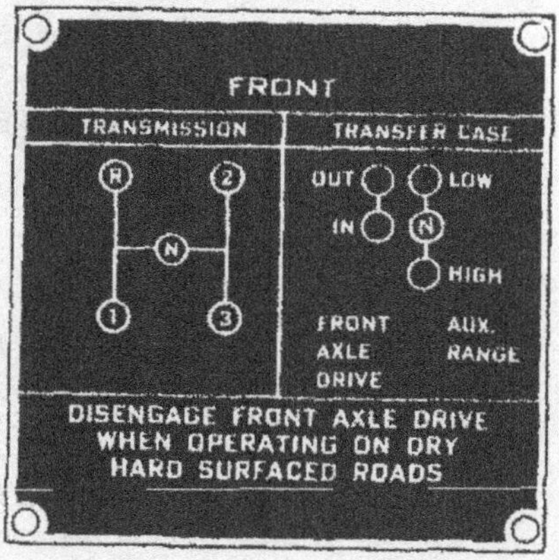

RA PD 305161

Figure 8—Shift Plate

is an electrical unit which indicates the fuel level in the tank, and only registers while the ignition switch is turned on.

(3) OIL PRESSURE GAGE (fig. 5). The oil pressure gage located on the instrument panel indicates the oil pressure when the engine is running.

(4) SPEEDOMETER (fig. 5). The speedometer on the instrument panel indicates in miles per hour the speed at which the vehicle is being driven. The odometer (in upper part of speedometer face) registers the total number of miles the vehicle has been driven. A trip indicator (in lower part of speedometer face) gives distance covered on any trip. Set trip indicator by turning the knurled control shaft extending through back of the speedometer.

(5) TEMPERATURE GAGE (fig. 5). The temperature gage registers the temperature of the solution in the cooling system.

1). Controls.

(1) BLACKOUT DRIVING LIGHT SWITCH (fig. 5). The blackout driving light switch (B.O. DRIVE) on the instrument panel controls the blackout driving light located on the left front fender, to furnish additional light during blackout periods. To operate light, first pull the blackout *light* switch button to the first position, then pull blackout *driving* light switch knob. To switch off the light, push in blackout *driving* light switch knob.

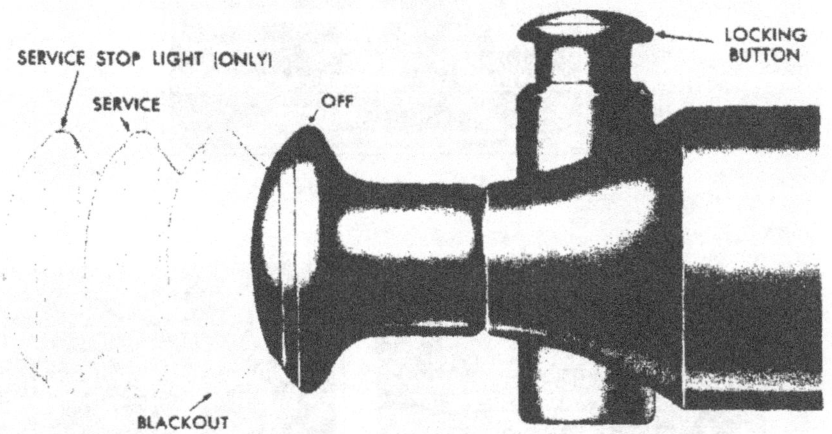

RA PD 645S6

Fimurm 9—Blmckovt Light Switch Operating Positions

(2) BLACKOUT LIGHT SWITCH (fig. 5). The knob on the instrument panel (LIGHTS) controls the entire lighting system, including the instrument panel lights, blackout driving light, and stop lights. A circuit-breaker type fuse, on the back of the switch, opens when a short circuit occurs, and closes when the thermostatic element cools. The light switch is a four-position push-pull type with a safety lock (fig. 9). When the control knob is pulled out to the first position, the blackout headlights and blackout stop and taillights are turned on.

The switch control knob travel is automatically locked in this position by the lock-out button to prevent accidentally turning on of the service (bright) lights in a blackout area. To obtain service lights, push in on lock-out control button on the left side of the switch, and pull out control knob to second position. When switch is in this position service headlights, service stop and taillights are turned on, and the panel lights can be turned on by pulling out on the knob (PANEL LIGHTS). CAUTION: *When driving during the day, press in lock-out control button, and pull control knob out to the last or stop light position to cause only the regular stop light to function.*

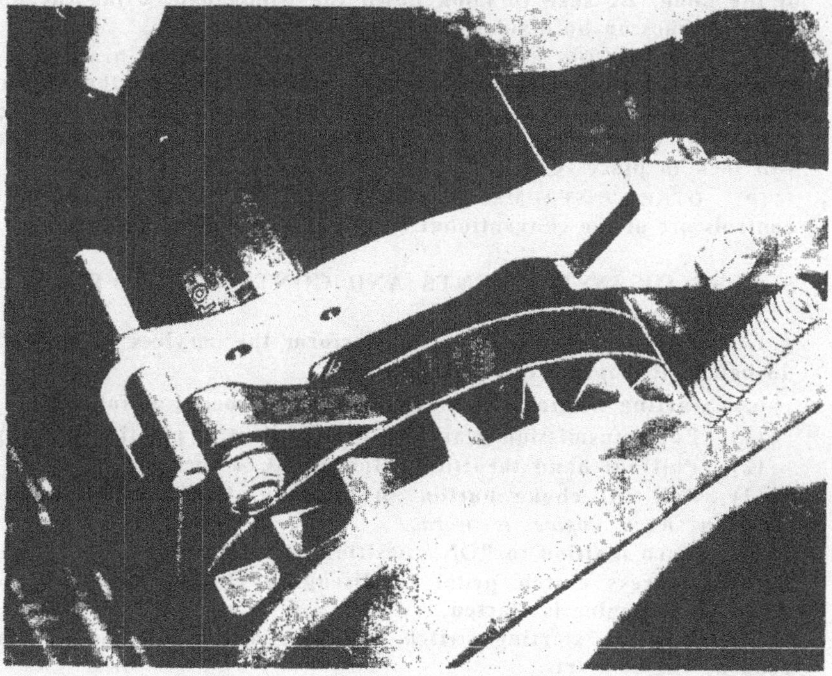

RA PD 305165

Figure 10—Generator Brace

(3) PANEL LIGHT SWITCH (fig. 5). The panel light switch knob (PANEL LIGHTS), located on the instrument panel, controls the lights to illuminate the panel instruments and controls. The blackout light switch (subpar. b (2) above) must be in service (bright light) position for this switch to control the panel lights.

(4) FIRE EXTINGUISHER (fig. 5). The fire extinguisher is mounted inside the left cowl panel. To remove, pull outward on the clamp release lever. To operate extinguisher, hold body in one hand and with the other, turn handle to left one-quarter turn, which releases plunger lock. Use pumping action to force liquid on base of fire. Read instructions on fire extinguisher plate.

(5) HAND BRAKE (fig. 5). The hand brake is applied by pulling out on the handle at the center of the instrument panel. Pull the handle out in a vertical position when the vehicle is parked. The brake is released by turning the handle one-quarter turn.

(6) WINDSHIELD ADJUSTING ARMS (fig. 5). The windshield adjustment arms are mounted on each end of the windshield frame. To open windshield, loosen knobs and push forward on lower part, then set by tightening the knobs.

(7) WINDSHIELD CLAMPS (fig. 5). The windshield clamps are located on the lower part of the windshield. Pull up on both clamps and unhook them, after which the windshield can be lowered on top of the hood. Be sure to hook down the windshield, using the hold-down catches on both sides of the hood.

(8) GENERATOR BRACE (fig. 10). The generator brace can be pulled up to release tension on the fan belt and stop the fan from throwing water over the engine when crossing a stream. Pull generator out to running position as soon as possible thereafter, and it will lock in place. CAUTION: *Be sure fan belt is on pulleys.*

(9) OTHER INSTRUMENTS AND CONTROLS. Other instruments and controls are of the conventional type, and are shown in figure 5.

5. USE OF INSTRUMENTS AND CONTROLS IN VEHICULAR OPERATION.

a. Before-operation Service. Perform the services in paragraph 13 before attempting to start the engine.

b. Starting Engine. To start the engine proceed as follows:

(1) Put transmission gearshift lever in neutral position (fig. 8).

(2) Pull out hand throttle button about % inch to 1 inch.

(3) Pull out choke button all the way. NOTE: *Choking is not necessary when engine is warm.*

(4) Turn ignition to "ON" position.

(5) Depress clutch pedal to disengage clutch, and hold pedal down while engine is started.

(6) Step on starting switch to crank again. Release switch as soon as engine starts.

(7) Adjust choke and throttle control buttons to obtain proper idling speed. As engine warms up, push choke button all the way in.

(8) Check oil pressure gage reading; at idle speed the indicator hand should show at least 10 on the gage.

(9) Check ammeter for charge reading. Check fuel gage for indication of fuel supply.

(10) After engine has operated a few minutes, check temperature gage reading. Normal operating temperature is between 160°F and 185°F.

(11) In extremely cold weather refer to paragraph 7.

c. Placing Vehicle in Motion.

(1) For daytime driving turn on service stop light (par. 4 b (2)).

(2) Place transfer case right-hand shift lever in rear position to

engage "HIGH" range, then place center shift lever in forward position to disengage front axle (fig. 8).

(3) Depress clutch pedal, and move transmission shift lever toward driver and backward to engage low (1st) gear (fig. 8).

(4) Release parking (hand) brake.

(5) Slightly depress accelerator to increase engine speed, and at the same time slowly release clutch pedal, increasing pressure on accelerator as clutch engages and vehicle starts to move. NOTE: *During the following operations perform procedures outlined in paragraph 14.*

(6) Increase speed to approximately 10 miles per hour, depress clutch pedal, and at the same time release pressure on accelerator. Move transmission shift lever out of low gear into neutral, and then into second gear. No double clutching is required. Release clutch pedal and accelerate engine.

(7) After vehicle has attained a speed of approximately 20 miles per hour, follow the same procedure as outlined above in order to shift into high (3rd) gear, moving the gearshift lever straight back.

d. Shifting to Lower Gears in Transmission. Shift to a lower gear before engine begins to labor, as follows: Depress clutch pedal quickly, shift to next lower gear, increase engine speed, release clutch pedal slowly, and accelerate. When shifting to a lower gear at any rate of vehicle speed, make sure that the engine speed is synchronized with vehicle speed before clutch is engaged.

e. Shifting Gears in Transfer Case (fig. 8). The transfer case is the means by which power is applied to the front and rear axles. In addition, the low gear provided by the transfer case further increases the number of speeds provided by the transmission. The selection of gear ratios depends upon the road and load conditions. Shift gears in the transfer case in accordance with the shift plate (fig. 8), and observe the instructions on the caution plate (fig. 7). The transmission gearshift does not in any way affect the selection or shifting of the transfer case gears. Vehicle may be driven by rear axle, or by both front and rear axles. The front axle cannot be driven independently.

(1) FRONT AXLE ENGAGEMENT. Front axle should be engaged only in off-the-road operation, slippery roads, steep grades, or during hard pulling. Disengage front axle when operating on average roads under normal conditions.

(a) Engaging Front Axle with Transfer Case in "HIGH" Range. With transfer case in "HIGH" range, move front axle drive shift lever to "IN" position. Depressing the clutch pedal will facilitate shifting.

(b) Disengaging Front Axle with Transfer Case in "HIGH" Range. Move front axle drive shift lever to "OUT" position. Depress the clutch pedal to facilitate shifting.

(c) Disengaging Front Axle when Transfer Case is in "LOW."

1. Depress clutch pedal, then shift transfer case lever into "HIGH."

2. Shift front axle drive lever into "OUT" position.

3. Release clutch pedal and accelerate engine to desired speed.

(2) ENGAGING TRANSFER CASE LOW RANGE. Transfer case LOW range cannot be engaged until front axle drive is engaged.

(a) Engage front axle drive (subpar. e (1) above).

(b) Depress clutch pedal and move transfer case shift lever into "N" (neutral) position.

(c) Release clutch pedal and accelerate engine.

(d) Depress clutch pedal again and move transfer case shift lever forward into "LOW" position.

(e) Release clutch pedal, and accelerate engine to desired speed.

(3) ENGAGING TRANSFER CASE-"LOW" to "HIGH." This shift can be made regardless of vehicle speed.

(a) Depress clutch pedal and move transfer case shift lever into "HIGH" position.

(b) Release clutch pedal, and accelerate engine to desired speed.

f. Stopping the Vehicle. Remove foot from accelerator, and apply brakes by depressing brake pedal.

(1) When vehicle speed has been reduced to engine idle speed, depress clutch pedal and move transmission shift lever to "N" (neutral) position (fig. 8).

(2) When vehicle has come to a complete stop, apply parking (hand) brake, and release clutch and brake pedals.

g. Reversing the Vehicle. To shift into reverse speed, first bring the vehicle to a complete stop.

(1) Depress clutch pedal.

(2) Move transmission shift lever to the left and forward into "R" (reverse) position.

(3) Release clutch pedal slowly, and accelerate as load is picked up.

h. Stopping the Engine. To stop the engine turn the ignition switch to "OFF" position. NOTE: *Before a new or reconditioned vehicle is first put into service, make run-in tests as outlined in section 10.*

6. TOWING THE VEHICLE.

a. Attaching Tow Line. To tow vehicle attach the chain, rope or cable to the front bumper bar at the frame side rail gusset (fig. 11). Do not tow from the middle of the bumper. To attach tow line, loop chain, rope, or cable over top of bumper, bring tow line up across front of bumper, and back on opposite side of frame, then hook or tie.

b. Towing to Start Vehicle. Place transfer case (aux. RANGE) shift lever of towed vehicle to the rear ("HIGH"). Place front axle drive shift lever in "OUT" (forward) position. Depress clutch pedal and engage transmission in high (3rd) speed. Switch ignition "ON," pull out choke control knob (if engine is cold), pull out throttle knob about 1 inch, release parking (hand) brake, and tow vehicle. After vehicle is under way, release clutch pedal slowly. As engine starts, regulate choke and throttle controls and disengage clutch, being careful to avoid overrunning towing vehicle or tow line.

c. **Towing Disabled Vehicle.** When towing a disabled vehicle exercise care so that no additional damage will occur.

(1) ALL WHEELS ON GROUND.

(a) If transfer case is *not* damaged, shift transmission and transfer case into neutral position and follow steps *(c)* and *(d)* below.

(b) If transfer case *is* damaged, disconnect both propeller shafts at the front and rear axles by removing the universal joint U-bolts, being careful not to lose the bearing races and rollers. Securely fasten the shafts to the frame with wire or remove dust cap and pull apart at the universal joint splines. Place bolts, nuts, rollers, and races in the glove compartment.

Figure 11—Chain Tow

(c) If the front axle differential or propeller shaft is damaged, remove front axle shaft driving flanges. Place front axle drive shift lever in "OUT" (forward) position and drive vehicle under own power.

(d) If the rear axle differential is damaged, remove the rear axle shafts; remove rear propeller shaft at rear universal joint U-bolts and front universal joint snap rings in forward flange, then drive out bearing cups. Place front axle drive shift lever in "IN" (rear) position and this will allow front axle drive to propel vehicle under own power.

(e) If rear propeller shaft only is damaged, remove as described in step *(d)* above.

(2) TOWING VEHICLE WITH FRONT OR REAR WHEELS OFF GROUND. If vehicle is to be towed in this manner be sure that transfer case shift lever is placed in "N" (neutral) position and front axle drive shift lever is placed in "OUT" (disengaged) position.

Section IV

OPERATION UNDER UNUSUAL CONDITIONS

7. OPERATION IN COLD WEATHER.

a. Purpose. Operation of automotive equipment at subzero temperatures presents problems that demand special precautions and extra careful servicing from both operation and maintenance personnel, if poor performance and total functional failure are to be avoided.

b. Gasoline. Winter grade of gasoline is designed to reduce cold weather starting difficulties; therefore, the winter grade motor fuel should be used in cold weather operation.

c. Storage and Handling of Gasoline. Due to condensation of moisture from the air, water will accumulate in tanks, drums, and containers. At low temperatures, this water will form ice crystals that will clog fuel lines and carburetor jets, unless the following precautions are taken:

(1) Strain the fuel through filter paper, or any other type of strainer that will prevent the passage of water. CAUTION: *Gasoline flowing over a surface generates static electricity that will result in a spark, unless means are provided to ground the electricity. Always provide a metallic contact between the container and the tank, to assure an effective ground.*

(2) Keep tank full, if possible. The more fuel there is in the tank, the smaller will be the volume of air from which moisture can be condensed.

(3) Add *y,* pint of denatured alcohol, Grade 3, to the fuel tank each time it is filled. This will reduce the hazard of ice formation in the fuel.

(4) Be sure that all containers are thoroughly clean and free from rust before storing fuel in them.

(5) If possible, after filling or moving a container, allow the fuel to settle before filling fuel tank from it.

(6) Keep all closures of containers tight to prevent snow, ice, dirt, and other foreign matter from entering.

(7) Wipe all snow or ice from dispensing equipment and from around fuel tank filler cap before removing cap to refuel vehicle.

d. Lubrication.

(1) TRANSMISSION AND DIFFERENTIAL.

(a) Universal gear lubricant, SAE 80, where specified on figure 14, is suitable for use at temperatures as low as —20°F. If consistent temperature below 0°F is anticipated, drain the gear cases while warm, and refill with Grade 75 universal gear lubricant, which is suitable for operation at all temperatures below +32°F. If Grade 75 universal gear lubricant is not available, SAE 80 universal gear lubricant diluted with the fuel used by the engine, in the proportion of one part fuel to six parts universal gear lubricant, may be used. Dilute make-up oil in the same proportion before it is added to gear cases.

(b) After engine has been warmed up, engage clutch, and maintain engine speed at fast idle for 5 minutes, or until gears can be engaged. Put transmission in low (first) gear, and drive vehicle for 100 yards, being careful not to stall engine. This will heat gear lubricants to the point where normal operation can be expected.

(2) CHASSIS POINTS. Lubricate chassis points with general purpose grease, No. 0.

(3) STEERING GEAR HOUSING. Drain housing, if possible, or use suction gun to remove as much lubricant as possible. Refill with universal gear lubricant, Grade 75, or, if not available, SAE 80 universal gear lubricant diluted with fuel used in the engine, in the proportion of one part fuel to six parts SAE 80 universal gear lubricant. Dilute make-up oil in the same proportion before it is added to the housing.

(4) OILCAN POINTS. For oilcan points where engine oil is prescribed for above 0°F, use light lubricating, preservative oil.

e. Protection of Cooling Systems.

(1) USE ANTIFREEZE COMPOUND. Protect the system with antifreeze compound (ethylene-glycol type) for operation below +32°F. The following instructions apply to use of new antifreeze compound.

(2) CLEAN COOLING SYSTEM. Before adding antifreeze compound, clean the cooling system, and completely free it from rust. If the cooling system has been cleaned recently, it may be necessary only to drain, refill with clean water, and again drain. Otherwise the system should be cleaned with cleaning compound.

(3) REPAIR LEAKS. Inspect all hoses, and replace if deteriorated. Inspect all hose clamps, plugs, and pet cocks and tighten if necessary. Repair all radiator leaks before adding antifreeze compound. Correct all leakage of exhaust gas or air into the cooling system.

(4) ADD ANTIFREEZE COMPOUND. When the cooling system is clean and tight, fill the system with water to about one-third capacity. Then add antifreeze compound, using the proportion of antifreeze compound to the cooling system capacity indicated below. Protect the system to at least 10°F below the lowest temperature expected to be experienced during the winter season.

ANTIFREEZE COMPOUND CHART
(for 11-quart capacity cooling system)

Temperature	Antifreeze Compound (ethylene-glycol type)
+10°F	3 qt
0°F	3¾ qt
−10°F	4½ qt
−20°F	4¾ qt
−30°F	5½ qt
−40°F	6 qt

(5) WARM THE ENGINE. After adding antifreeze compound, fill with water to slightly below the filler neck; then start and warm the engine to normal operating temperature.

(6) TEST STRENGTH OF SOLUTION. Stop the engine and check the solution with a hydrometer, adding antifreeze compound if required.

(7) INSPECT WEEKLY. In service, inspect the coolant weekly for strength and color. If rusty, drain and clean cooling system thoroughly, and add new solution of the required strength.

(8) CAUTIONS.

(a) Antifreeze compound is the only antifreeze material authorized for ordnance materiel.

(b) It is essential that antifreeze solutions be kept clean. Use only containers and water that are free from dirt, rust, and oil.

(c) Use an accurate hydrometer. To test a hydrometer, use one part antifreeze compound to two parts water. This solution will produce a hydrometer reading of 0°F.

(d) Do not spill antifreeze compound on painted surfaces.

f. **Electrical Systems.**

(1) **GENERATOR AND CRANKING MOTOR. Check the brushes, commutators, and bearings. See that the commutators are clean. The large surges of current which occur when starting a cold engine require good contact between brushes and commutators.**

(2) **WIRING. Check, clean, and tighten all connections, especially the battery terminals. Care should be taken that no short circuits are present.**

(3) **COIL. Check coil for proper functioning by noting quality of spark.**

(4) **DISTRIBUTOR. Clean thoroughly, and clean or replace points. Check the points frequently. In cold weather, slightly pitted points may prevent engine from starting.**

(5) **SPARK PLUGS. Clean and adjust or replace, if necessary. If it is difficult to make the engine fire, reduce the gap to 0.00S inch less than that recommended for normal operation (par. 67 l»). This will make ignition more effective at reduced voltages likely to prevail.**

(6) TIMING. Check carefully. Care should be taken that the spark is not unduly advanced nor retarded.

(7) BATTERY.

(a) The efficiency of batteries decreases sharply with decreasing temperatures, and becomes practically nil at —40°F. Do not try to start the engine with the battery when it has been chilled to temperatures below —30°F until battery has been heated, unless a warm slave battery is available. See that the battery is always fully charged, with the hydrometer reading between 1.275 and 1.300. A fully charged battery will not freeze at temperatures likely to be encountered even in arctic climates, but a fully discharged battery will freeze and rupture at +5°F.

(b) Do not add water to a battery when it has been exposed to subzero temperatures unless the battery is to be charged immediately. If water is added and the battery not put on charge, the layer of water will stay at the top and freeze before it has a chance to mix with the acid.

(8) LIGHTS. Inspect the lights carefully. Check for short circuits and presence of moisture around sockets.

(9) ICE. Before every start, see that the spark plugs, wiring, or other electrical equipment is free from ice.

g. Starting and Operating Engine.

(1) INSPECT CRANKING MOTOR MECHANISM. Be sure that no heavy grease or dirt has been left on the cranking motor throwout mechanism. Heavy grease or dirt is liable to keep the gears from being meshed, or cause them to remain in mesh after the engine starts running. The latter will ruin the cranking motor and necessitate repairs.

(2) USE OF CHOKE. A full choke is necessary to secure the rich air-fuel mixture required for cold weather starting. Check the butterfly valve to see that it closes all the way, and otherwise functions properly.

(3) CARBURETOR AND FUEL PUMP. The carburetor, which will give no appreciable trouble at normal temperatures, is liable not to operate satisfactorily at low temperatures. Be sure the fuel pump has no leaky valves or diaphragm, as this will prevent the fuel pump from delivering the amount of fuel required to start the engine at low temperatures, when turning speeds are reduced to 30 to 60 revolutions per minute.

(4) AIR CLEANERS. At temperatures below 0°F do not use oil in air cleaners. The oil will congeal and prevent the easy flow of air. Wash screens in dry-cleaning solvent, dry, and replace. Ice and frost formations on the air cleaner screens can cause an abnormally high intake vacuum in the carburetor air horn hose, resulting in collapse.

(5) FUEL SYSTEM. Remove and clean sediment bulb, strainers, etc., daily. Also drain fuel tank sump daily to remove water and dirt.

(6) STARTING THE ENGINE. Observe the following precautions in addition to the normal starting procedure (par. 5 a and b).

(a) Clean ignition wires and outside of spark plugs of dirt and frost.

(b) Free distributor point arm on post and clean points.

(c) Be sure carburetor choke closes fully.

(d) Operate fuel pump hand lever to fill carburetor bowl (fig. 12).

(e) Free up engine with hand crank or use slave battery.

(7) Stop engine if no oil pressure shows on gage.

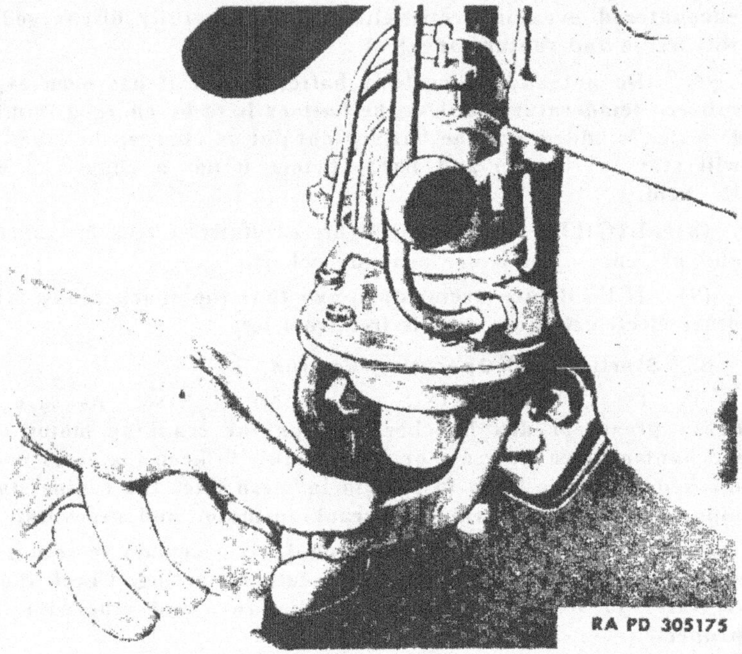

Figure 12—Fuel Pump, Hand Operation

(g) Engage clutch to warm up transmission oil before attempting to move vehicle.

(h) Check engine operation for proper condition (par. 13 b (22)).

h. Chassis.

(1) BRAKE BANDS. Brake bands, particularly on new vehicles, have a tendency to bind when they are very cold. Always have a blowtorch handy to warm up these parts, if they bind prior to moving, or attempting to move, the vehicle. Parking the vehicle with the brake released will eliminate most of the binding. Precaution must be taken, under these circumstances, to block the wheels or otherwise prevent movement of the vehicle.

(2) EFFECT OF LOW TEMPERATURES ON METALS. Inspect the vehicle frequently. Shock resistance of metals, or resistance against breaking, is greatly reduced at extremely low temperatures. Operation of vehicles on hard, frozen ground causes strain and jolting which will result in screws breaking, or nuts jarring loose.

(3) SPEEDOMETER CABLE. Disconnect the oil-lubricated speedometer cable at the drive end when operating the vehicle at temperatures of —30°F and below. The cable will often fail to work properly at these temperatures, and sometimes will break, due to the excessive drag caused by the high viscosity of the oil with which it is lubricated.

8. OPERATION IN HOT WEATHER.

A. PROTECTION OF VEHICLE. In extremely hot weather avoid the continuous use of low gear ratios whenever possible. Check and replenish oil and water frequently. If a flooded condition of the engine is experienced in starting, pull the throttle control out, push choke control in, and use the cranking motor. When engine starts, adjust throttle control.

(1) COOLING SYSTEM. Rust formation occurs more rapidly during high temperatures; therefore, add rust preventive solution to the cooling system, or clean and flush the system at frequent intervals.

(2) LUBRICATION. Lubricate the vehicle for hot weather operation (par. 8).

(3) ELECTRICAL SYSTEM. Check the battery solution level frequently during hot weather operation, anct add water as required to keep it above the top of the plates. If hard starting is experienced in hot, damp weather or quick changes in temperature, dry the spark plugs, wires, and both inside and outside of distributor cap.

9. OPERATION IN SAND.

A. OPERATION. Reduce tire pressures in desert terrain if character of sand demands this precaution. When operating in sand deep enough to cause the use of a lower gear, do not exceed the speed specified on the caution plate for the particular gear ratio (fig. 7).

B. STARTING THE VEHICLE. When starting the vehicle in sand, gravel, or soft terrain, engage the front wheel drive (par. 5 E (1)). Release clutch pedal slowly so the wheels will not spin and "dig in," necessitating a tow or "winch-out."

C. CLUTCH. Do not attempt to "jump" or "rock" the vehicle out with a quick engagement of the clutch, particularly if a tow or winch is available. Racing the engine usually causes the wheels to "dig in" farther.

D. AIR CLEANER. In sandy territory clean the carburetor air cleaner more often. The frequency of cleaning depends upon the severity of the sandy condition.

e. Radiator. In desert operation check the radiator coolant supply frequently, and see that the air passages of the core do not become clogged.

f. For additional information on technique of operating the vehicle in sand, refer to FM 31-25.

10. OPERATION IN LANDING.

a. Inspection. As soon as possible after completing a landing or operation in water, inspect the vehicle for water in the various units.

(1) ENGINE. Drain the engine crankcase oil. If water or sludge is found, flush the engine, using a mixture of half engine oil SAE 10 and half kerosene. Before putting in new oil, clean the valve chamber, drain and clean the oil filter, and install a new filter element.

(2) FUEL SYSTEM. Inspect the carburetor bowl, fuel strainers, fuel pump, filter, fuel tank, and lines. Clean the air cleaner and change the oil.

(3) POWER TRAIN. Inspect the front and rear axle housings, wheel bearings, transmission, and transfer case lubricant for presence of sludge. If sludge is found, renew the lubricant after cleaning the units with a mixture of half engine oil SAE 10 and half kerosene. Lubricate the propeller shaft universal joints and spring shackles to force out any water which might damage parts.

11. DECONTAMINATION.

a. Protection. For protective measures against chemical attacks and decontamination refer to FM 17-59.

Section V

FIRST ECHELON PREVENTIVE MAINTENANCE SERVICE

12. PURPOSE.

a. To ensure mechanical efficiency it is necessary that the vehicle be systematically inspected at intervals each day it is operated, also weekly, so that defects may be discovered and corrected before they result in serious damage or failure. Certain scheduled maintenance services will be performed at these designated intervals. The services set forth in this section are those performed by driver or crew before operation, during operation, at halt, and after operation and weekly.

b. Driver preventive maintenance services are listed on the back of "Driver's Trip Ticket and Preventive Maintenance Service Record," W.D. Form No. 48, to cover vehicles of all types and models. Items peculiar to specific vehicles, but not listed on W.D. Form No. 48, are covered in manual procedures under the items to which they are related. Certain items listed on the form that do not pertain to the vehicle involved are eliminated from the procedures as written into the manual. Every organization must thoroughly school each driver in performing the maintenance procedures set forth in manuals, whether they are listed specifically on W.D. Form No. 48 or not.

c. The items listed on W.D. Form No. 48 that apply to this vehicle are expanded in this manual to provide specific procedures for accomplishment of the inspections and services. These services

driver, and are not necessarily in the same numerical order as shown on W.D. Form No. 48. The item numbers, however, are identical with those shown on that form.

d. The general inspection of each item applies also to any supporting member or connection, and generally includes a check to see whether the item is in good condition, correctly assembled, secure, or excessively worn.

(1) The inspection for "good condition" is usually an external visual inspection to determine whether the unit is damaged beyond safe or serviceable limits. The term "good condition" is explained further by the following: not bent or twisted, not chafed or burned, not broken or cracked, not bare or frayed, not dented or collapsed, not torn or cut.

(2) The inspection of a unit to see that it is "correctly assembled" is usually an external visual inspection to see whether or not it is in its normal assembled position in the vehicle.

(3) The inspection of a unit to determine if it is "secure" is usually an external visual examination, a hand-feel, wrench, or pry-bar check for looseness. Such an inspection should include any brackets, lock washers, lock nuts, locking wires, or cotter pins used in assembly.

(4) "Excessively worn" will be understood to mean worn, close to or beyond, serviceable limits, and likely to result in failure if not replaced before the next scheduled inspection.

e. Any defects or unsatisfactory operating characteristics beyond the scope of the first echelon to correct must be reported at the earliest opportunity to the designated individual in authority.

13. BEFORE-OPERATION SERVICE.

a. This inspection schedule is designed primarily as a check to see that the vehicle has not been tampered with or sabotaged since the After-operation Service was performed. Various combat conditions may have rendered the vehicle unsafe for operation, and it is the duty of the driver to determine whether or not the vehicle is in condition to carry out any mission to which it is assigned. This operation will not be entirely omitted, even in extreme tactical situations.

b. Procedures. Before-operation Service consists of inspecting items listed below according to the procedure described, and correcting or reporting any deficiencies. Upon completion of the service, results should be reported promptly to the designated individual in authority.

(1) ITEM 1, TAMPERING AND DAMAGE. Examine exterior of vehicle, engine, wheels, brakes, and steering control for damage by falling debris, shell fire, sabotage, or collision. If wet, dry the ignition parts to ensure easy starting.

(2) ITEM 2, FIRE EXTINGUISHER. Be sure fire extinguisher is full, nozzle is clean, and mountings secure.

(3) ITEM 3, FUEL, OIL, AND WATER. Check fuel tank, crankcase, and radiator for leaks or tampering. Add fuel, oil, or water as needed. Have value of antifreeze checked. If, during period when antifreeze is used, it becomes necessary to replenish a considerable amount of water, report unusual losses.

(4) ITEM 4, ACCESSORIES AND DRIVES. Inspect carburetor, generator, regulator, cranking motor, and water pump for loose connections and security of mountings. Inspect carburetor and water pump for leaks.

(5) ITEM 6, LEAKS, GENERAL. Look on ground under vehicle for indications of fuel, oil, water, brake fluid, or gear oil leaks. Trace leaks to source, and correct or report to higher authority.

(6) ITEM 7, ENGINE WARM-UP. Start engine, observe cranking motor action, listen for unusual noise, and note cranking speed. Idle engine only fast enough to run smoothly. Proceed immediately with following services while engine is warming up.

(7) ITEM 8, CHOKE. AS engine warms, push in choke as required for smooth operation, and to prevent oil dilution.

(8) ITEM 9, INSTRUMENTS.

(a) *Fuel Gage.* Fuel gage should indicate approximate amount of fuel in tank.

(b) *Oil Pressure Gage.* Normal oil pressure should not be below 10 with engine idling, and should range from 40 to 50 at running speeds (at normal operating temperature). If gage fails to register within 30 seconds, stop engine, and correct or report to higher authority.

(c) *Temperature Indicator.* Temperature should rise slowly during warm-up. Normal operating temperature range is 160°F to 185°F.

(d) *Ammeter.* Ammeter should show high charge for short period after starting and positive (plus) reading above 12 to 15 miles per hour with lights and accessories off. Zero reading is normal with lights and accessories on.

(9) ITEM 10, HORN AND WINDSHIELD WIPERS. Sound horn, tactical situation permitting, for proper operation and tone. Check both wipers for secure attachment and normal full contact operation through full stroke.

(10) ITEM 11, GLASS AND REAR VIEW MIRROR. Clean windshield and rear view mirror and inspect for cracked, discolored, or broken glass. Adjust mirror.

(11) ITEM 12, LIGHTS AND REFLECTORS. Try switches in each position and see if lights respond. Lights and warning reflectors must be securely mounted, clean, and in good condition. Test foot control of headlight beams.

(12) ITEM 13, WHEEL AND FLANGE NUTS. Observe whether or not all wheel and flange nuts are present and tight.

(13) ITEM 14, TIRES. If time permits, test tires with gage, including spare; normal pressure is 35 pounds with tires cold. Inspect tread and carcass for cuts and bruises. Remove imbedded objects from treads.

(14) ITEM 15, SPRINGS AND SUSPENSION. Inspect springs for sagged or broken leaves, shifted leaves, and loose or missing rebound clips.

(15) ITEM 16, STEERING LINKAGE. Examine steering gear case, connecting links, and Pitman arm for security and good condition. Test steering adjustment, and free motion of steering wheel.

(16) ITEM 17, FENDERS AND BUMPERS. Examine fenders and bumpers for secure mounting and serviceable condition.

FIRST ECHELON PREVENTIVE MAINTENANCE SERVICE

(17) ITEM 18, TOWING CONNECTIONS. Examine pintle hook for secure mounting and serviceable condition. Be sure pintle latches properly and locks securely.

(18) ITEM 19, BODY AND LOAD. Examine body and load (if any) for damage. Be sure there is a cap on front drain hole under fuel tank. See that rear drain hole cap is available in glove compartment. CAUTION: *Rear drain hole cap should be installed when about to pass through deep water.*

(19) ITEM 20, DECONTAMINATOR. Examine decontaminator for full charge and secure mountings.

(20) ITEM 21, TOOLS AND EQUIPMENT. See that tools and equipment are all present, properly stowed, and serviceable.

(21) ITEM 23, DRIVER'S PERMIT AND FORM 26. Driver must have his operator's permit on his person. See that vehicle manuals, Lubrication Guide, Form No. 26 (accident report) and W.D. AGO Form No. 478 (MWO and Major Unit Assembly Replacement Record) are present, legible, and properly stowed.

(22) ITEM 22, ENGINE OPERATION. Accelerate engine and observe for unusual noises indicating compression or exhaust leaks; worn, damaged, loose, and inadequately lubricated parts or misfiring.

(23) ITEM 25, DURING-OPERATION SERVICE. Begin the During-operation Service immediately after the vehicle is put in motion.

14. DURIING-OPERATION service.

a. While vehicle is in motion, listen for any sounds such as rattles, knocks, squeals, or hums that may indicate trouble. Look for indications of trouble in cooling system, and smoke from any part of the vehicle. Be on the alert to detect any odor of overheated components or units such as generator, brakes, or clutch; check for fuel vapor from a leak in fuel system, exhaust gas, or other signs of trouble. Any time the brakes are used, gears shifted, or vehicle turned, consider this a test and notice any unsatisfactory or unusual performance. Watch the instruments frequently. Notice promptly any unusual instrument indication that may signify possible trouble in system to which the instrument applies.

b. Procedures. During-operation Service consists of observing items listed below according to the procedures following each item, and investigating any indications of serious trouble. Notice minor deficiencies to be corrected or reported at earliest opportunity, usually at next scheduled halt.

(1) ITEM 27, FOOT AND HAND BRAKES. Foot brakes must stop vehicle smoothly without side pull and within reasonable distance. There should be at least \'- reserve brake pedal travel and *y,-mch* free travel. Hand brake must securely hold vehicle on reasonable incline with % reserve ratchet travel. There must be %-inch clearance (on cable) between relay crank and lower end of hand brake conduit.

¼-TON 4 x 4 TRUCK (WILLYS-OVERLAND MODEL MB and FORD MODEL GPW)

(2) ITEM 28, CLUTCH. Clutch must operate smoothly without chatter, grabbing, or slipping. Free clutch pedal travel of three-quarter inch is normal.

(3) ITEM 29, TRANSMISSION. Gearshift mechanism must operate smoothly, and not creep out of mesh.

(4) ITEM 29, TRANSFER CASE. Gearshift mechanism must operate smoothly and not creep out Of mesh.

(5) ITEM 31, ENGINE AND CONTROLS. Observe whether or not engine responds to controls, and has maximum pulling power without unusual noises, stalling, misfiring, overheating or unusual exhaust smoke. If radio noise is reported during operation of the vehicle, the driver will cooperate with the radio operator in locating the interference. See paragraph 178.

(6) ITEM 32, INSTRUMENTS. During operation observe the readings of all instruments frequently to see if they are indicating properly.

(a) Fuel Gage. Fuel gage must register approximate amount of fuel in tank.

(b) Oil Pressure Gage. Oil pressure gage should register 10 with engine running idle, and 40 to 50 at operating speeds.

(c) Temperature Indicator. Temperature indicator should show a temperature of 160°F to 185°F after warm-up under normal conditions.

(d) Speedometer. Speedometer should show speed of vehicle without noise or fluctuation of indicator needle. Odometer should register accumulating trip and total mileage.

(e) Ammeter. Ammeter should show zero reading with lights on, zero or positive (plus) charge with lights off, and slightly higher positive (plus) charge for short time immediately after starting.

(7) ITEM 33, STEERING GEAR. Observe steering for excessive pulling of vehicle to either side, wandering, or shimmy.

(8) ITEM 34, CHASSIS. Listen for unusual noises from wheel or axles.

(9) ITEM 35, BODY. Observe body for sagging springs, loose or torn top or windshield cover, if in use.

15. AT-HALT SERVICE.

a. At-halt Service may be regarded as the minimum maintenance procedure, and should be performed under all tactical conditions, even though more extensive maintenance services must be slighted or omitted altogether.

b. Procedures. At-halt Service consists of investigating any deficiencies noted during operation, inspecting items listed below according to the procedures following the items, and correcting any deficiencies found. Deficiencies not corrected should be reported promptly to the designated individual in authority.

FIRST ECHELON PREVENTIVE MAINTENANCE SERVICE

(1) ITEM 38, FUEL, OIL AND WATER. Check fuel supply, oil, and coolant; add, as required, for complete operation of vehicle to the next refueling point. If, during period when antifreeze is used, an abnormal amount of water is required to refill radiator, have coolant tested with hydrometer, and add antifreeze if required.

(2) ITEM 39, TEMPERATURES. Feel each brake drum and wheel hub, transmission, transfer case, and front and rear axles for overheating. Examine gear cases for excessive oil leaks.

(3) ITEM 40, AXLE AND TRANSFER CASE VENTS. Observe whether axle and transfer case vents are present, and see that they are not damaged or clogged.

(4) ITEM 41, PROPELLER SHAFT. Inspect propeller shaft for looseness, damage, or oil leaks.

(5) ITEM 42, SPRINGS. Look for broken spring leaves or loose clips and U-bolts.

(6) ITEM 43, STEERING LINKAGE. Examine steering control mechanism and linkage for damage or looseness. Investigate any irregularities noted during operation.

(7) ITEM 44, WHEEL AND FLANGE NUTS. Observe whether or not all wheel and axle flange nuts are present and tight.

(8) ITEM 45, TIRES. Inspect tires, including spare, for fiats or damage, and for cuts or foreign material imbedded in tread.

(9) ITEM 46, LEAKS, GENERAL. Check around engine and on ground beneath the vehicle for excessive leaks. Trace to source, and correct cause or report to higher authority.

(10) ITEM 47, ACCESSORIES AND BELTS. See that fan, water pump and generator are securely mounted, that fan belt is adjusted to 1-inch deflection, and is not badly frayed. If radio noise during operation of the engine was observed, examine all radio noise suppression capacitors, at coil, ignition and starting switches, generator, regulator, and radio terminal box; suppressors at spark plugs and distributor, and all bond straps for damage, and loose mountings or connections.

(11) ITEM 48, AIR CLEANER. If dusty or sandy conditions have been encountered, examine oil sump for excessive dirt. Service if required. CAUTION: *Do not apply oil to element after cleaning.*

(12) ITEM 49, FENDERS AND BUMPERS. Inspect fenders and bumpers for looseness or damage.

(13) ITEM 50, TOWING CONNECTIONS. Inspect pintle hook and trailer light socket for serviceability.

(14) ITEM 51, BODY LOAD AND TARPAULIN. Inspect vehicle and trailed vehicle loads for shifting; see that tarpaulins are properly secured and not damaged.

(15) ITEM 52, APPEARANCE AND GLASS. Clean windshield, mirror, light lenses, and inspect vehicle for damage.

16. AFTER-OPERATION AND WEEKLY SERVICE.

a. After-operation Service is particularly important because at this time the driver inspects his vehicle to detect any deficiencies that may have developed, and corrects those he is permitted to handle. He should report promptly, to the designated individual in authority, the results of his inspection. If this schedule is performed thoroughly, the vehicle should be ready to roll again on short notice. The Before-operation Service, with a few exceptions, is then necessary only to ascertain whether the vehicle is in the same condition in which it was left upon completion of the After-operation Service. The After-operation Service should never be entirely omitted, even in extreme tactical situations, but may be reduced, if necessary, to the bare fundamental services outlined for the At-halt Service.

b. Procedures. When performing the After-operation Service the driver must remember and consider any irregularities noticed during the day in the Before-operation, During-operation, and At-halt Services. The After-operation Service consists of inspecting and servicing the following items. Those items of the After-operation Service that are marked by an asterisk (*) require additional Weekly Service, the procedures for which are indicated in step *(b)* of each applicable item.

(1) ITEM 54, FUEL, OIL, AND WATER. Check coolant and oil levels, and add as needed. Fill fuel tank. Refill spare cans. During period when antifreeze is used, have hydrometer test made of coolant if loss from boiling or other cause has been considerable. Add antifreeze with water if required.

(2) ITEM 55, ENGINE OPERATION. Listen for miss, backfire, noise, or vibration that might indicate worn parts, loose mountings, faulty fuel mixture, or faulty ignition.

(3) ITEM 56, INSTRUMENTS. Inspect all instruments to see that they are securely connected, and not damaged.

(4) ITEM 57, HORN AND WINDSHIELD WIPERS. Test horn for sound, if tactical situation permits. See that horn is securely mounted and properly connected. Operate both windshield wipers. See that blades contact the glass effectively throughout full stroke.

(5) ITEM 58, GLASS AND REAR VIEW MIRROR. Clean glass of windshield and rear view mirror. Examine for secure mounting and damage.

(6) ITEM 59, LIGHTS AND REFLECTORS. Observe whether or not lights operate properly with the switch in "ON" positions, and go out when switch is off. See that stop light operates properly. Clean lenses and warning reflectors.

(7) ITEM 60, FIRE EXTINGUISHER. Be sure fire extinguisher is full, nozzle is clean, and that extinguisher is mounted securely.

(8) ITEM 61, DECONTAMINATOR. Examine decontaminator for good condition and secure mounting.

(9) ITEM 62, *BATTERY.

(a) See that battery is clean, securely mounted, and not leaking. Inspect electrolyte level, which should be *y,* inch above plates with caps in place and vents open. Clean cables as required.

(b) *Weekly.* Clean top of battery. Remove battery caps, and add water to *y,* inch above plates. (Use distilled water if available; if not use clean, drinkable water.) CAUTION: *Do not overfill.* Clean posts and terminals if corroded, and apply light coat of grease. Tighten terminals as needed. Tighten hold-down assembly. Clean battery carrier if corroded.

(10) ITEM 63, *ACCESSORIES AND BELTS.

(a) Test fan belt for deflection of 1 inch. Examine belt for good condition; it must not be frayed. Timing hole cover must be closed and tightened.

(b) *Weekly.* Tighten all accessories such as carburetor, generator, regulator, cranking motor, fan, water pump, and hose connections; examine fan belt for fraying, wear, cracking, or presence of oil.

(11) ITEM 64, *ELECTRICAL WIRING.

(a) See that all ignition wiring and accessible low voltage wiring is in good condition, clean, correctly and securely assembled and mounted.

(b) *Weekly.* Tighten all loose wiring connections or electrical unit mountings. Pay particular attention to radio noise suppression units such as: capacitors, bond straps, and spark plug and distributor suppressors.

(12) ITEM 65, *AIR CLEANER.

(a) Examine oil in air cleaner oil cup to see that it is at proper level, and not excessively dirty. Clean element and refill oil cup as required. CAUTION: *Do not apply oil to element after cleaning.*

(b) *Weekly.* Remove, clean, and dry air cleaner element and oil cup. Fill cup to indicated oil level (approximately 5/g qt). Do not apply oil to element after cleaning.

(13) ITEM 66, *FUEL FILTERS.

(a) Examine fuel filter for leaks.

(b) *Weekly.* Remove plug from bottom of dash-mounted fuel filter. Allow water and sediment to drain out. Be sure plug is replaced tightly, and does not leak.

(14) ITEM 67, ENGINE CONTROLS. Examine engine controls for wear or disconnected linkage.

(15) ITEM 68, *TIRES.

(a) Inspect tires for cuts or abnormal tread wear; remove foreign bodies from tread; inflate to 35 pounds when tires are cold.

(b) *Weekly.* Replace badly worn or otherwise unserviceable tires.

(16) ITEM 69, *SPRINGS.

(a) Examine springs for sag, broken or shifted leaves, loose or missing rebound clips, or shackles.

(b) *Weekly.* Aline springs, and tighten U-bolts and shackles as required.

(17) ITEM 70, STEERING LINKAGE. Examine steering wheel column, gear case, Pitman arm, drag link, tie rod, and steering arm to see if they are bent, loose, or inadequately lubricated.

(18) ITEM 71, PROPELLER SHAFT. Inspect propeller shaft and universal joints for loose connections, lubrication leaks, or damage.

(19) ITEM 72, *AXLE AND TRANSFER VENTS.

(a) See that axle and transfer case vents are in good condition, clean, and secure.

(b) *Weekly.* Remove, clean, and replace vents.

(20) ITEM 73, LEAKS, GENERAL. Check under hood and beneath the vehicle for indications of fuel, oil, water, or brake fluid leaks.

(21) ITEM 74, GEAR OIL LEVELS. After units have cooled, inspect differential transmission and transfer unit lubricant levels. Lubricant should be level with bottom of filler hole. Observe gear cases for leaks.

(22) ITEM 76, FENDERS AND BUMPERS. Fenders and bumpers must be in good condition and secure.

(23) ITEM 77, *TOWING CONNECTIONS.

(a) Inspect pintle hook and towed-load connections for looseness or damage.

(b) *Weekly.* Tighten pintle hook mounting bolts, and lubricate pintle hook as required.

(24) ITEM 78, BODY AND TARPAULINS. Inspect body, top, and windshield cover for damage and proper stowage. Make sure rear drain below fuel tank is open, and that cap is in glove compartment.

(25) ITEM 82, *TIGHTEN.

(a) Tighten any loose wheel, axle drive flange, and spring U-bolt nuts.

(b) *Weekly.* Tighten all vehicle assembly or mounting nuts or screws that inspection indicates require tightening.

(26) ITEM 83, *LUBRICATE AS NEEDED.

(a) Lubricate spring shackles and steering linkage, if lubrication is needed.

(b) *Weekly.* Lubricate points indicated on current vehicle Lubrication Guide as requiring weekly attention, also points that experience and operating conditions indicate need lubrication. Observe latest lubrication directives.

(27) ITEM 84, *CLEAN ENGINE AND VEHICLE.

(a) Clean dirt and trash from inside of body. Keep sump under fuel tank cleaned of dirt and water. Remove excessive dirt or grease from exterior of the engine.

(b) *Weekly.* Wash vehicle if possible. If not possible, wipe off thoroughly; clean engine.

(28) ITEM 85, TOOLS AND EQUIPMENT. Check to see that all tools and equipment assigned to vehicle are present and secure.

SECTION VI

LUBRICATION

17. LUBRICATION GUIDE.

a. War Department Lubrication Guide No. 501 (figs. 13 and 14) prescribes lubrication maintenance for the *y^-ton* 4 x 4 truck.

b. A Lubrication Guide is placed on or is issued with each vehicle and is to be carried with it at all times. In the event the vehicle is received without a Guide, the using arm shall immediately requisition a replacement from the Commanding Officer, Fort Wayne Ordnance Depot, Detroit 32, Mich.

c. Lubrication instructions on the Guide are binding on all echelons of maintenance and there shall be no deviations from these instructions.

d. Service intervals specified on the Guide are for normal operation conditions. Reduce these intervals under extreme conditions such as excessively high or low temperatures, prolonged periods of high speed, continued operation in sand or dust, immersion in water, or exposure to moisture, any one of which may quickly destroy the protective qualities of the lubricant and require servicing in order to prevent malfunctioning or damage to the materiel.

e. Lubricants are prescribed in the "Key" in accordance with three temperature ranges; above +32°F, +32°F to 0°F, and below 0°F. Determine the time to change grades of lubricants by maintaining a close check on operation of the vehicle during the approach to change-over periods. Be particularly observant when starting the engine. Sluggish starting is an indication of thickened lubricants and the signal to change to grades prescribed for the next lower temperature range. Ordinarily it will be necessary to change grades of lubricants *only when air temperatures are consistently in the next higher or lower range,* unless malfunctioning occurs sooner due to lubricants being too thin or too heavy.

18. DETAILED LUBRICATION INSTRUCTIONS.

a. Lubrication Equipment. Each piece of materiel is supplied with lubrication equipment adequate to maintain the materiel. Be sure to clean this equipment both before and after use. Operate lubricating guns carefully and in such manner as to insure a proper distribution of the lubricant.

b. Points of Application.

(1) Red circles surrounding lubrication fittings, grease cups, oilers and oil holes make them readily identifiable on the vehicle. Wipe clean such lubricators and the surrounding surface before lubricant is applied.

No. 501

WAR DEPARTMENT LUBRICATION GUIDE

ORDNANCE DEPARTMENT

TRUCK, 1/4 TON, 4x4 (FORD-WILLYS)

SNL G-503

For detailed instructions, refer to TM.

TABLE OF CAPACITIES AND LUBRICANTS TO BE USED

UNIT	CAPACITY (Approx.)	LOWEST EXPECTED AIR TEMPERATURE		
		+32° F. and above	+32° F. to 0° F.	Below 0° F.
Crankcase (Including Oil Filter)	5 qt.	OE SAE 30	OE SAE 10	Refer to OFSB 6-11
Transmission	3/4 qt.	GO SAE 90	GO SAE 80	GO Grade 75
Transfer Case	1 1/2 qt.			
Differentials (each)	1 1/4 qt.			

CAUTION Lubricate Dotted Arrow Points on BOTH SIDES. Points on OPPOSITE SIDE are indicated by Dotted Short-Shaft Arrows.

NOTE—See Reverse Side for lubrication of TRAILER

Serviced From Engine Compartment

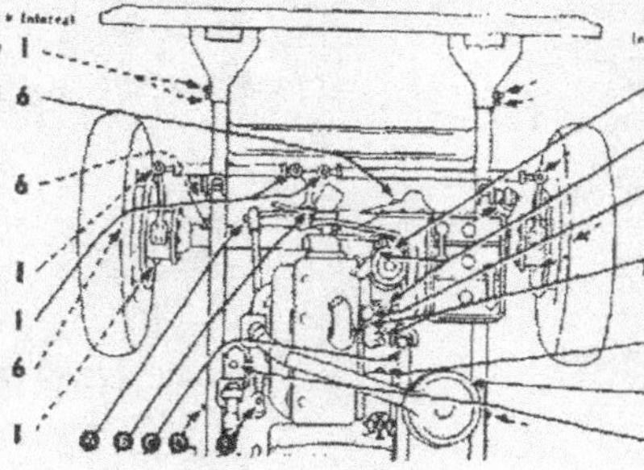

Lubricant • Interval

- Spring Shackle CG 1
- Front Axle Differential GO 6 — Drain and refill; Check level weekly (Note 7)
- Shock Absorbers SA 6 — (Some models) (Notes 15 and 16)
- Tie Rod CG 1
- Tie Rods (Inner) CG 1
- Front Wheel Bearings WB 6 — Remove, clean and repack
- Universal Joint and Steering CG 1

Interval • Lubricant

- 1 *Oil Filter* — Drain (Note 6) — C, fig.
- 1 OE Generator — (Early models) (Note 15) — A, fig.
- 1 OE Crankcase (See Table) — Drain and refill (Note 5); Check level daily — B, fig.
- 1 OE Distributor Shaft
- 6 OE Distributor — (Note 10) — E, fig.
- 1 OE Cranking Motor — 6 to 8 drops — F, fig.
- D OE Air Cleaner — Check level (Note 4) — D, fig.
- 1 GO Steering Gear — F, fig.

RA PD 3051

Knuckle Bearings (Note 8)

Drag Link CG 1

Steering Bellcrank CG 1

Universal Joint CG 1 (Note 9)

Spring Bolt CG 1

Drag Link CG 1

Clutch and Brake Pedals CG 1

Transmission GO 6 Drain and refill (Early WILLYS Models, right side) Check level weekly (Note 7)

Spring Bolt CG 1

Rear Wheel Bearings WB 6 Remove, clean and repack

Shock Absorbers SA 6 (Some models) (Notes 15 and 16)

Rear Axle Differential GO 6 Drain and refill Check level weekly (Note 7)

Spring Shackle CG 1

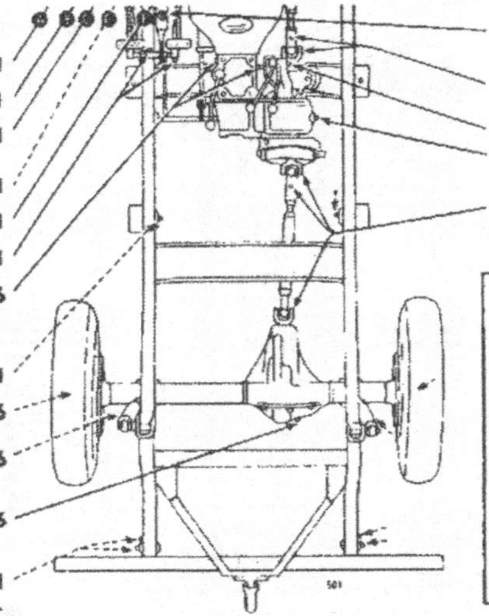

1 HB Brake Master Cylinder (Remove cover on toeboard) Fill to ½ in. from top — B, fig. 19

1 CG Universal and Slip Joints (Note 9) — E, fig. 16; F, fig. 16

1 CG Trans. Case Shift Lever Shaft — D, fig. 16

6 GO Transfer Case Drain and refill Check level weekly (Note 7) — C, fig. 16

1 CG Universal and Slip Joints (Note 9) — E, fig. 16; F, fig. 16

KEY

Lubricants	Intervals
OE—OIL, engine Except crankcase SAE 30 (above +32°F.) SAE 10 (+32°F. to 0°F.) PS (below 0°F.)	D—Daily
GO—LUBRICANT, gear, universal See Table	1—1,000 miles
CG—GREASE, general purpose No. 1 (above +32°F.) No. 0 (below +32°F.)	6—6,000 miles
WB—GREASE, general purpose, No. 2	Check Daily
HB—FLUID, brake, hydraulic	Crankcase
SA—FLUID, shock-absorber, light	Air Cleaner
PS—OIL, lubricating, preservative, special	

NOTES Additional Lubrication and Service Instructions on Individual Units and Parts NOTES

COLD WEATHER: For Lubrication and Service below 0°F., refer to OFSB 6-11.

1. FITTINGS—Clean before applying lubricant. Lubricate until new lubricant is forced from the bearing, unless otherwise specified. CAUTION: Lubricate chassis points after washing truck and trailer.

2. INTERVALS indicated are for normal service. For extreme conditions of speed, heat, water, sand, mud, snow, rough roads, dust, etc., reduce interval by ⅓ or ½, or more if conditions warrant.

3. CLEANING — SOLVENT, dry-cleaning, or OIL, fuel, diesel, will be used to clean or wash all parts. Use of gasoline for this purpose is prohibited. All parts will be thoroughly dry before relubrication.

4. AIR CLEANER—Daily, check level and refill oil reservoir to bead level with used crankcase oil or OE. Every 1,000 miles, daily under extreme dust conditions, remove and wash all parts. From 0°F. to −40°F., use SA. Below −40°F., remove oil and operate dry.

5. CRANKCASE—Drain only when engine is

(Notes continued on Reverse Side)

No. 501 (NOT TO BE REPRODUCED in whole or in part without permission of the Office of the Chief of Ordnance.) CHEK-CHART

24 Jan 44
Supersedes all previous issues

Figure 13—Lubrication Guide—Truck, 1/4-Ton, 4 x 4 (Ford-Willys)

RA PD 305160B

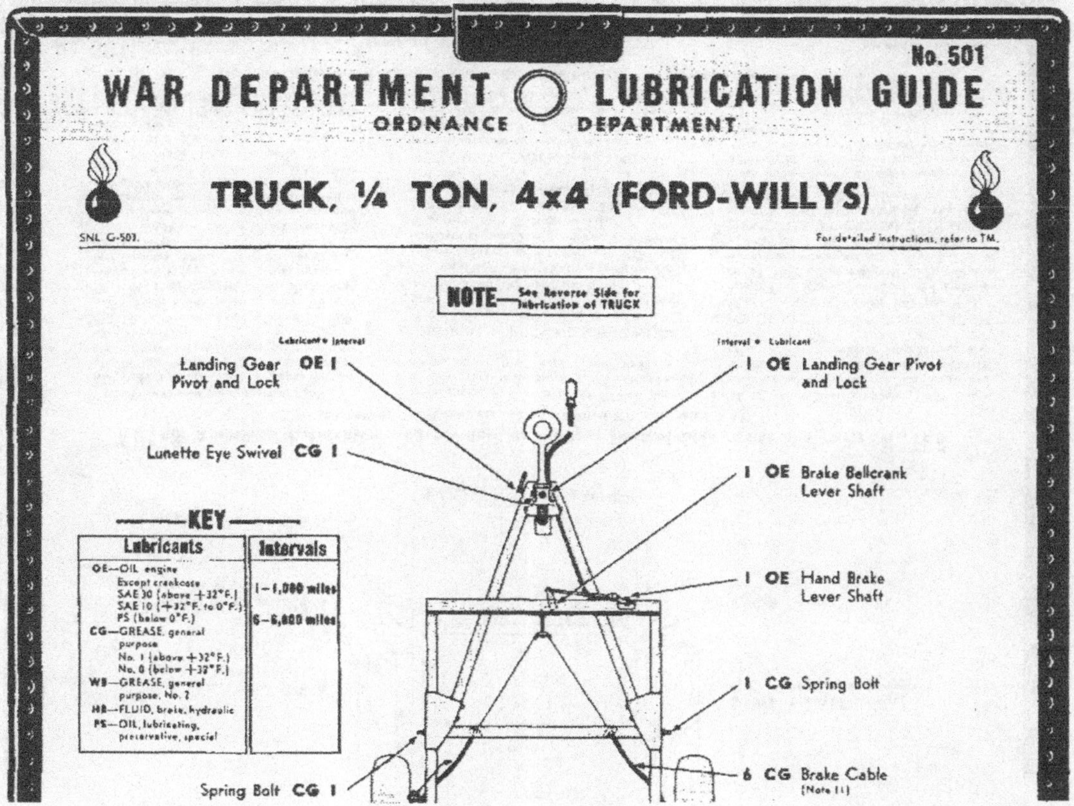
No. 501
WAR DEPARTMENT LUBRICATION GUIDE
ORDNANCE DEPARTMENT
TRUCK, 1/4 TON, 4x4 (FORD-WILLYS)
SNL G-503.
For detailed instructions, refer to TM.
NOTE—See Reverse Side for lubrication of TRUCK
Lubricant • Interval
Landing Gear Pivot and Lock OE 1
Lunette Eye Swivel CG 1
Interval • Lubricant
1 OE Landing Gear Pivot and Lock
1 OE Brake Bellcrank Lever Shaft
1 OE Hand Brake Lever Shaft
1 CG Spring Bolt
6 CG Brake Cable (Note 11)
Spring Bolt CG 1
KEY
Lubricants
OE—OIL, engine
Except crankcase
SAE 30 (above +32°F.)
SAE 10 (+32°F. to 0°F.)
PS (below 0°F.)
CG—GREASE, general purpose
No. 1 (above +32°F.)
No. 0 (below +32°F.)
WB—GREASE, general purpose, No. 2
HB—FLUID, brake, hydraulic
PS—OIL, lubricating, preservative, special
Intervals
1—1,000 miles
6—6,000 miles
RA PD 330851

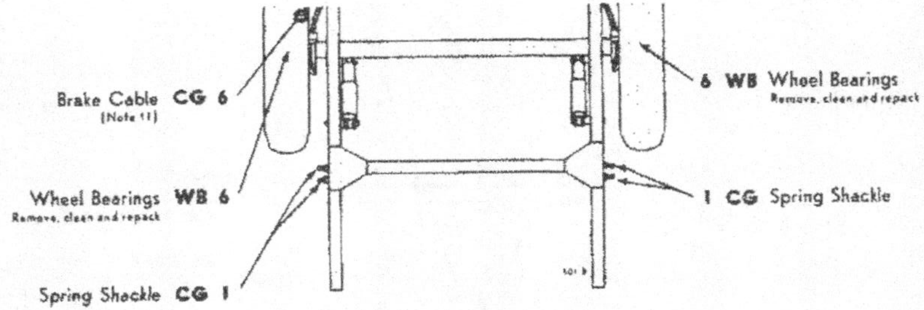

TRAILER, 1/4 TON, 2-WHEEL

NOTES Additional Lubrication and Service Instructions on Individual Units and Parts NOTES

COLD WEATHER: For Lubrication and Service below 0°F., refer to OFSB 6-11.

(Notes continued from Reverse Side)

hot. Refill to FULL mark on gage. Run engine a few minutes and recheck oil level. CAUTION: Be sure pressure gage indicates oil is circulating.

6. OIL FILTER—Every 1,000 miles, drain sediment. Every 6,000 miles or more often if filter becomes clogged, drain filter, clean inside and renew element. Run engine a few minutes, check crankcase level, add OE to FULL mark on gage.

7. GEAR CASES—Weekly, check level with truck on level ground and, if necessary, add lubricant to plug level. Check only before operation while lubricant is cold. Drain and refill at intervals indicated on guide. Drain only after operation. On early WILLYS models, skid plate must be removed to reach drain plug.

8. UNIVERSAL JOINT AND STEERING KNUCKLE BEARINGS—Every 1,000 miles, remove plug at rear and add CG to level of filler plug hole. Every 6,000 miles, remove, clean, dry, inspect and refill to plug level.

9. UNIVERSAL JOINTS AND SLIP JOINTS—Apply CG to joints until it shows at joint cross and to slip joint until it shows at the end of spline. Use hand gun only.

10. DISTRIBUTOR—Every 6,000 miles, wipe distributor breaker cam lightly with CG and lubricate breaker arm pivot and wick under rotor with 1 to 2 drops of OE.

11. TRAILER BRAKE CABLES—Every 6,000 miles, slide cable conduit forward, clean and coat with CG.

12. SPEEDOMETER CABLE—Every 6,000 miles, remove core and coat lightly with CG No. 0.

13. RUBBER BUSHINGS—Every 1,000 miles, apply H8 to shock absorber linkage. CAUTION: Do not use oil.

14. OIL CAN POINTS—Every 1,000 miles, lubricate all carburetor, clutch, brake and throttle linkages, pintle hook, handbrake cable, trailer handbrake connections and linkage with OE.

15. POINTS REQUIRING NO LUBRICATION SERVICE—Clutch Release Bearing, Water Pump, Fan, Shock Absorbers and Linkage on all FORD and early WILLYS models, Generator (late models).

16. POINTS TO BE SERVICED AND/OR LUBRICATED BY ORDNANCE MAINTENANCE PERSONNEL—Shock Absorbers (late WILLYS). (Refer to TM.)

Copy of this Guide will be carried on the materiel at all times. These lubrication instructions are binding on all echelons of maintenance.

By Order of the Secretary of War:
G. C. Marshall, Chief of Staff

Requisition replacement guides from the Commanding Officer, Fort Wayne Ordnance Depot, Detroit, Michigan.

No. 501 [NOT TO BE REPRODUCED in whole or in part without permission of the Office of the Chief of Ordnance] CHEK-CHART

24 Jan 44
Supersedes all previous issues.

RA PD 330851B

Figure 14—Lubrication Guide—Trailer, 1/4-Ton, 2-wheel

¼-TON 4 x 4 TRUCK (WILLYS-OVERLAND MODEL MB and FORD MODEL GPW)

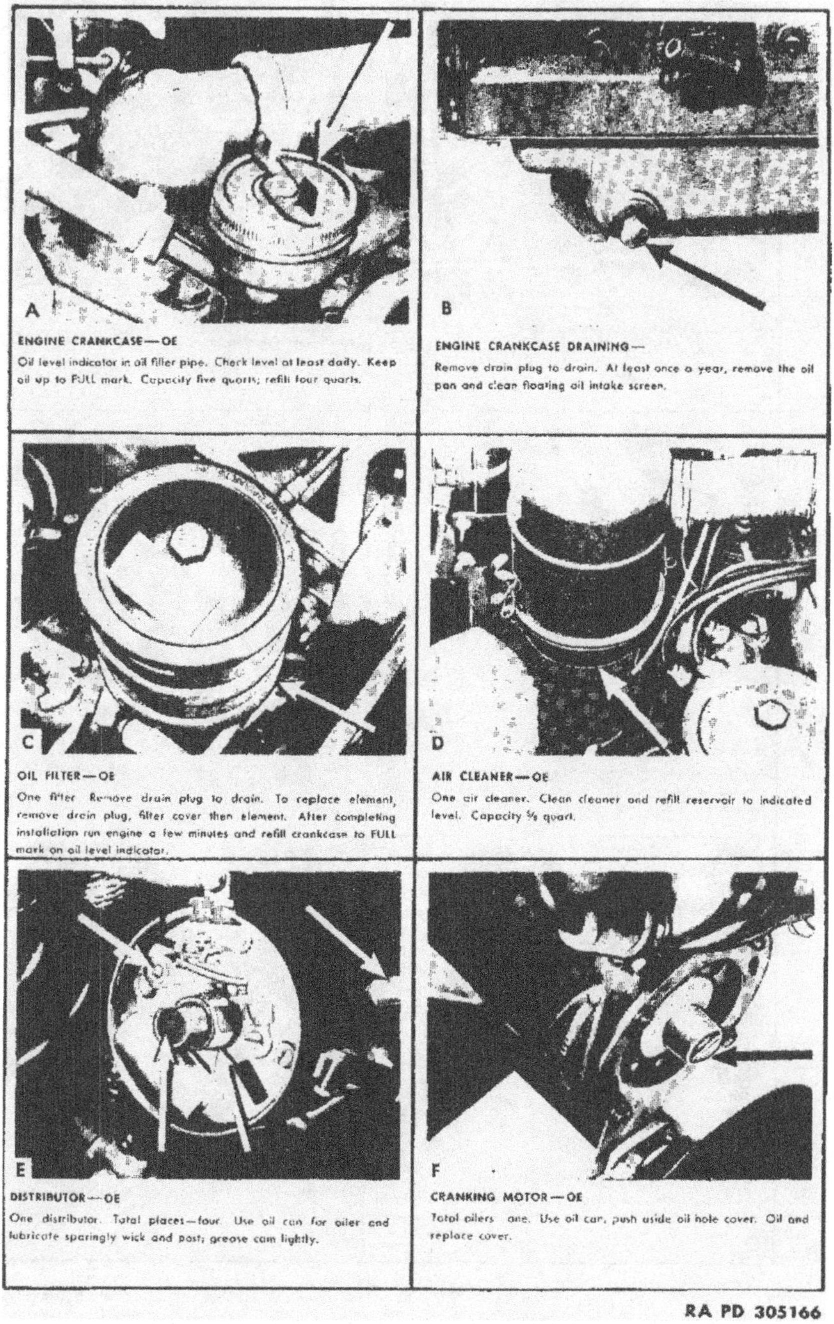

Figure 15—Engine Lubrication Points

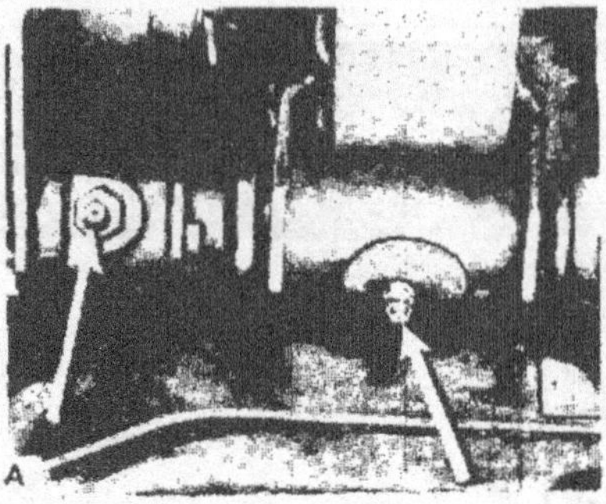

A

CLUTCH AND BRAKE PEDAL SHAFT—CG

One pedal shaft. Total fittings—two. Use pressure gun on fittings until grease shows.

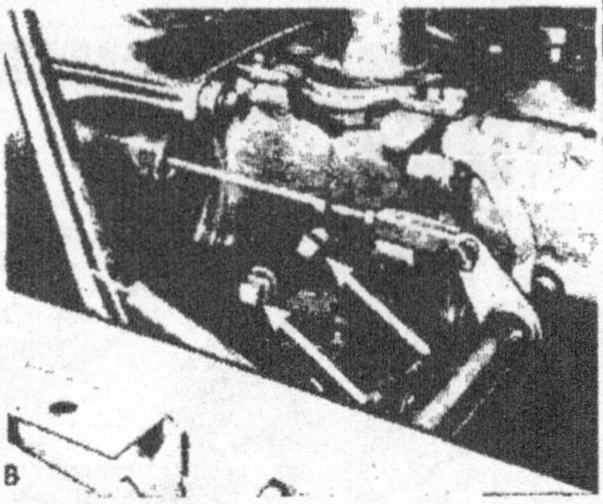

B

TRANSMISSION—GO

One transmission. Total plugs—two (filler and drain). Use gear oil pump. Drain and refill to bottom of filler plug hole. Capacity ¾ quart.

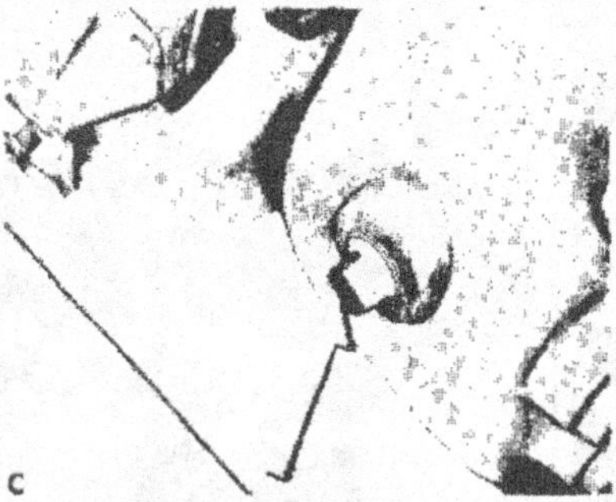

C

TRANSFER CASE—GO

One transfer case. Total plugs—two (filler and drain). Use gear oil pump. Drain and refill to bottom of filler plug hole. Capacity 1½ quarts.

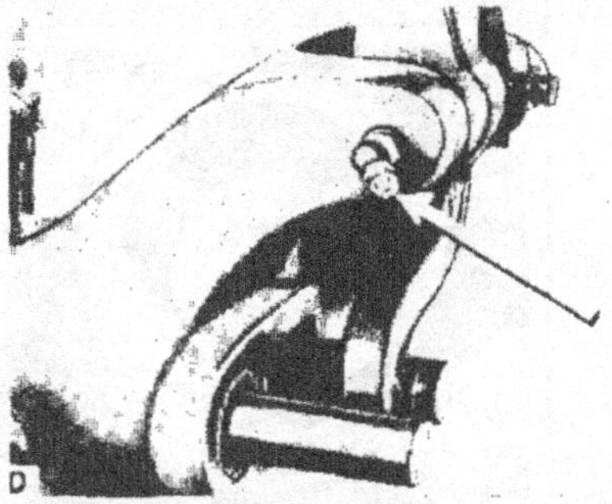

D

TRANSFER CASE SHIFT LEVER SHAFT—CG

One shift lever shaft. Total fittings—one. Use pressure gun on fitting until grease shows.

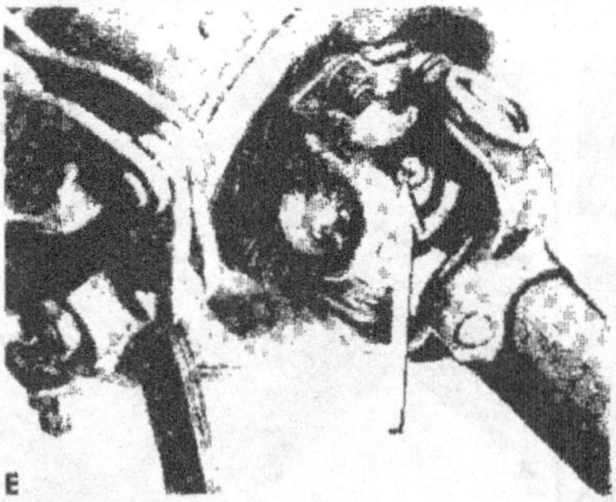

E

PROPELLER SHAFT UNIVERSAL JOINTS—CG

Four universal joints. Total fittings—four. Use pressure gun (hand) with adaptor. CAUTION: Do not use high pressure grease gun because of damage to seals.

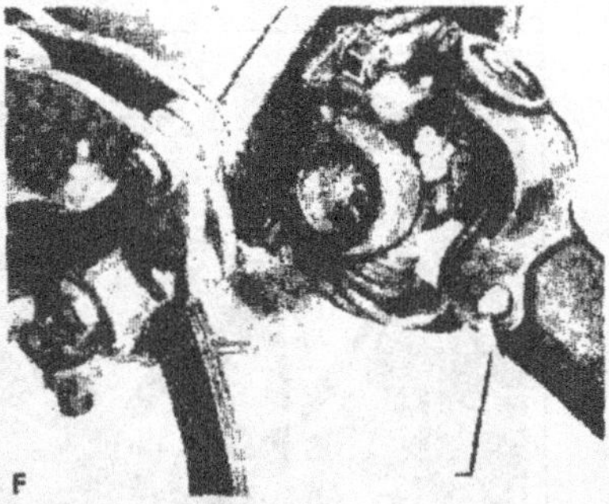

F

PROPELLER SHAFT SLIP JOINT—CG

Two slip joints. Total fittings—two. Use pressure gun on fittings until grease shows.

RA PD 305167

Figure 16—Pedal Shafts and Power Train Lubrication Points

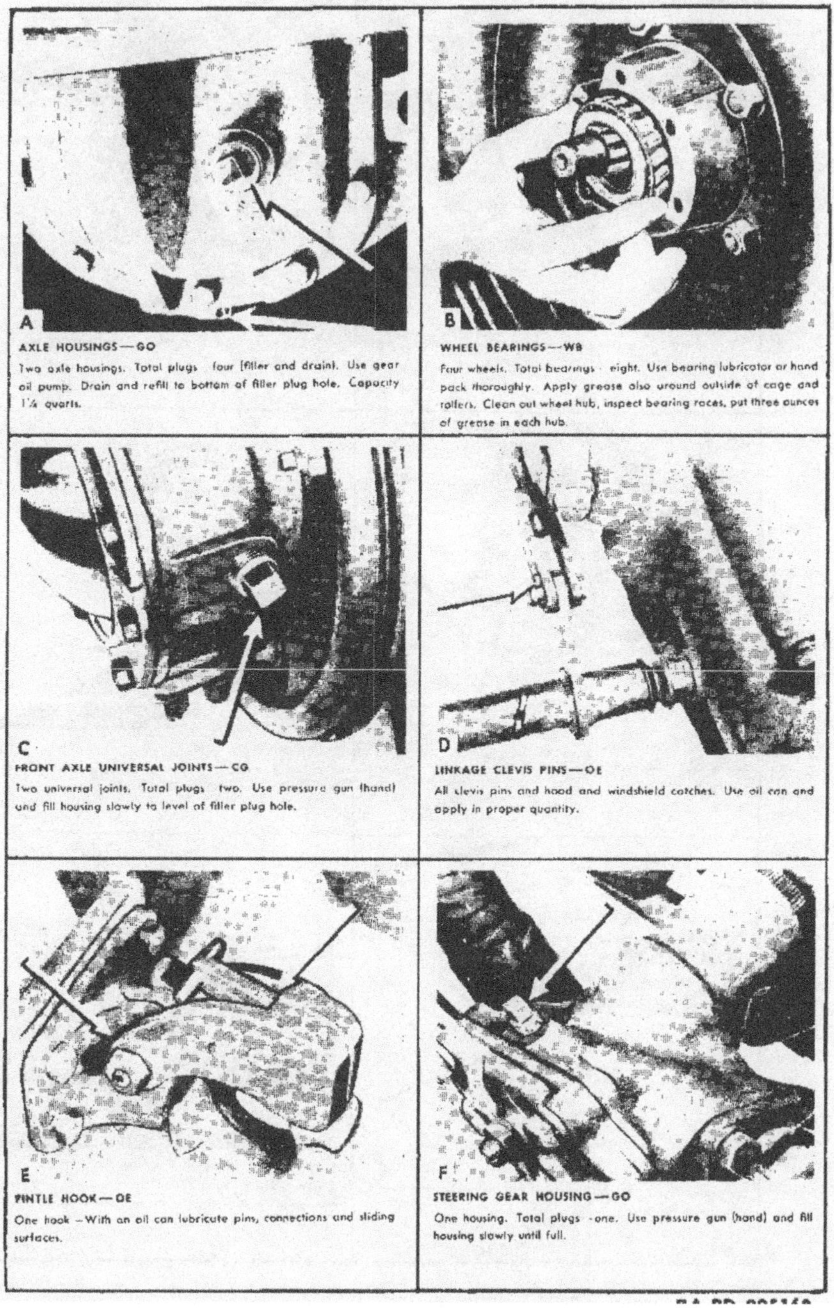

Figure 17—Axle, Wheel, Pintle, and Steering Gear Housing Lubrication Points

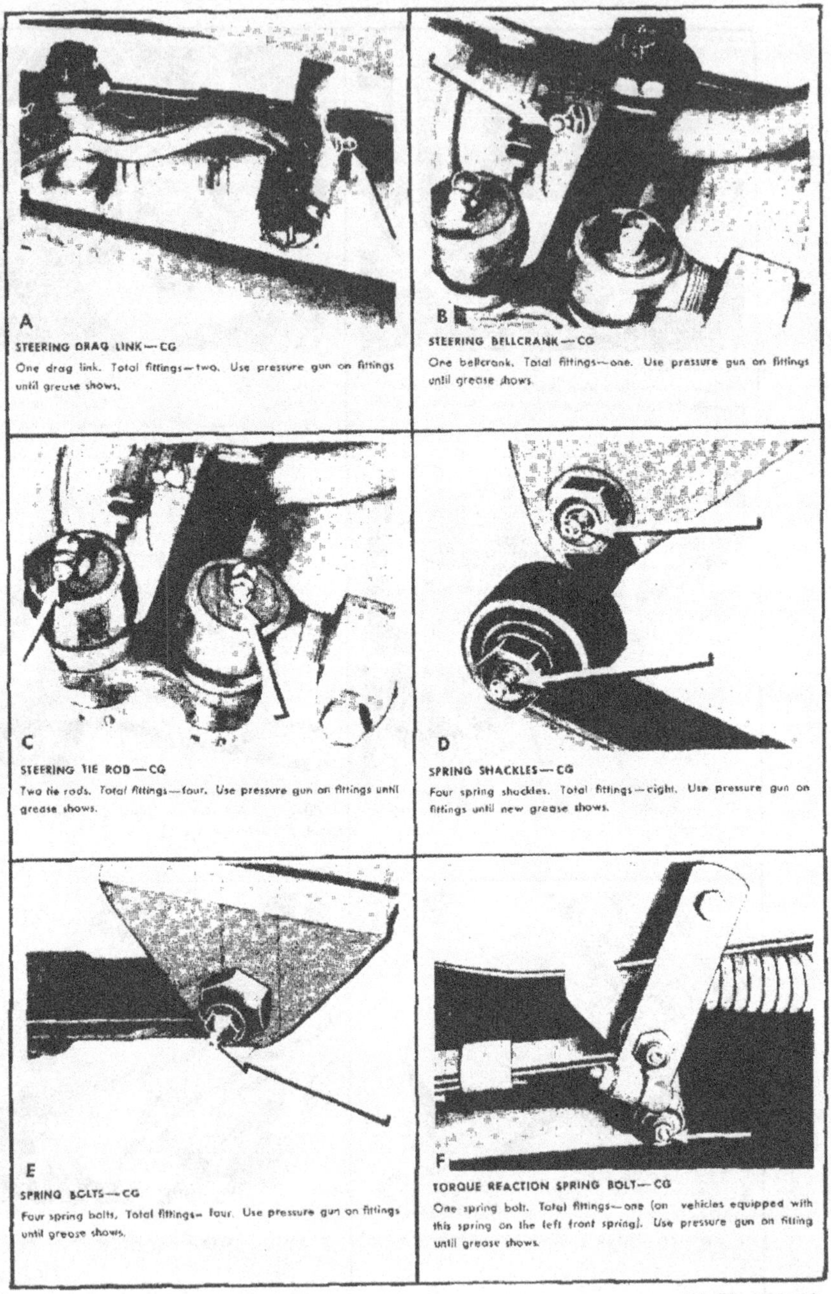

RA PD 305169

Figure 18—Steering Gear and Spring Lubrication Points

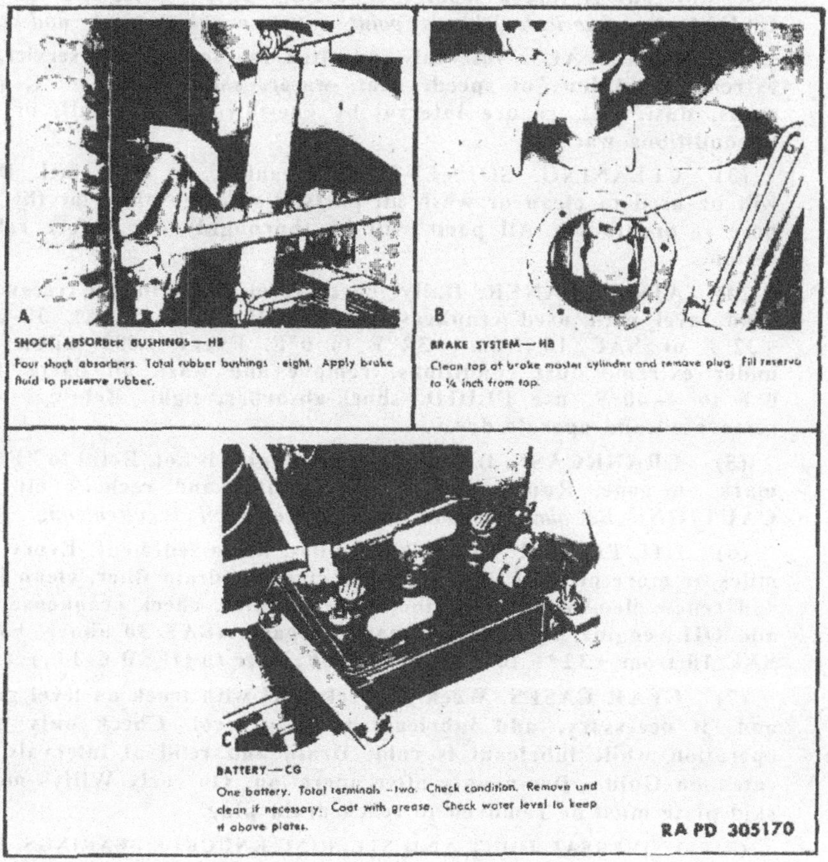

Figure 19—Shock Absorber, Masfer *Cylinder, and Battery Lubrication Points*

(2) Where relief valves are provided, apply new lubricant until the old lubricant is forced from the vent. Exceptions are specified in notes on the Lubrication Guide.

c. Cleaning. Use SOLVENT, dry-cleaning, or OIL, fuel, Diesel, to clean or wash all parts. Use of gasoline for this purpose is prohibited. After washing, dry all parts thoroughly before applying lubricant.

d. Lubrication Notes on Individual Units and Parts. The following instructions supplement those notes on the Lubrication Guide which pertain to lubrication and service of individual units and parts. All note references in the Guide itself are to the paragraph below having the corresponding number.

LUBRICATION

(1) FITTINGS. Clean before applying lubricant. Lubricate until new lubricant is forced from the bearing, unless otherwise specified. CAUTION: *Lubricate chassis points after washing truck and trailer.*

(2) INTERVALS. Intervals indicated are for normal service. For extreme conditions of speed, heat, water, sand, mud, snow, rough roads, dust, etc., reduce interval by one-third or one-half, or more if conditions warrant.

(3) CLEANING. SOLVENT, dry-cleaning, or OIL, fuel, Diesel, will be used to clean or wash all parts. Use of gasoline for this purpose is prohibited. All parts will be thoroughly dry before relubrication.

(4) AIR CLEANER. Daily, check level and refill oil reservoir to bead level with used crankcase oil or OIL, engine, SAE 30 above +32°F or SAE 10 from +32°F to 0°F. Every 1,000 miles, daily under extreme dust conditions, remove and wash all parts. From 0°F to —40°F, use FLUID, shock-absorber, light. Below —40°F, remove oil and operate dry.

(5) CRANKCASE. Drain only when engine is hot. Refill to "FULL" mark on gage. Run engine a few minutes and recheck oil level. CAUTION: *Be sure pressure gage indicates oil is circulating.*

(6) OIL FILTER. Every 1,000 miles, drain sediment. Every 6,000 miles or more often if filter becomes clogged, drain filter, clean inside and renew element. Run engine a few minutes, check crankcase level, add OIL, engine, to "FULL" mark on gage. (SAE 30 above +32°F; SAE 10 from +32°F to 0°F; below 0°F, refer to OFSB 6-11.)

(7) GEAR CASES. Weekly, check level with truck on level ground and, if necessary, add lubricant to plug level. Check only before operation while lubricant is cold. Drain and refill at intervals indicated on Guide. Drain only after operation. On early Willys models, skid plate must be removed to reach drain plug.

(8) UNIVERSAL JOINT AND STEERING KNUCKLE BEARINGS. Every 1,000 miles, remove plug at rear and add GREASE, general purpose, No. 1 above +32°F or No. 0 below +32°F, to level of filler plug hole. Every 6,000 miles, remove, clean, dry, inspect and refill to plug level.

(a) Remove brake tube and brake backing plate screws. This permits the removal of the axle spindle, the complete axle shaft, and the universal jo^nt assembly. Care should be taken not to injure the outer oil seal assembly in the housing.

(b) Wash the axle shaft and universal joint thoroughly in SOLVENT, dry-cleaning, and dry.

(c) Clean and repack upper and lower steering spindle bearings within the universal housing and reassemble entire unit.

(9) UNIVERSAL JOINTS AND SLIP JOINTS. Apply GREASE, general purpose, No. 1, above +32°F, or No. 0 below +32°F, to joints until it shows at joint cross, and to slip joint until it shows at the end of spline. Use hand gun only.

(10) DISTRIBUTOR. Every 6,000 miles, wipe distributor breaker cam lightly with GREASE, general purpose, No. 1, above +32°F or No. 0, below +32°F, and lubricate breaker arm pivot and wick under rotor with 1 to 2 drops of OIL, engine, SAE 30 above +32°F; SAE 10 from +32°F to 0°F; OIL, lubricating, preservative, special, below 0°F.

(11) TRAILER BRAKE CABLES. Every 6,000 miles, slide cable conduit forward, clean and coat with GREASE, general purpose, No. 1 above +32°F and No. 0 below +32°F.

(12) SPEEDOMETER CABLE. Every 6,000 miles, remove core and coat lightly with GREASE, general purpose, No. 0.

(13) RUBBER BUSHINGS. Every 1,000 miles, apply FLUID, brake, hydraulic, to shock absorber linkage. CAUTION: *Do not use oil.*

(14) OILCAN POINTS. Every 1,000 miles, lubricate all carburetor, clutch, brake and throttle linkages, pintle hook and hand brake cable with OIL, engine, SAE 30, above +32°F; SAE 10, +32°F to 0°F; OIL, lubricating, preservative, special, below 0°F.

(15) POINTS REQUIRING NO LUBRICATION SERVICE. These are the clutch release bearing, water pump, fan, shock absorbers and linkage on all Ford and early Willys models, generator (late models), speedometer cable.

(16) POINTS TO BE SERVICED AND/OR LUBRICATED BY ORDNANCE MAINTENANCE PERSONNEL ONLY. These are the shock absorbers (late Willys). Every 6,000 miles, remove and disassemble the shock absorbers. Unscrew linkage eye and refill with FLUID, shock-absorber, light.

(17) WHEEL BEARINGS. Remove bearing cone assemblies from hub and wash spindle and inside of hub. Inspect bearing races and replace if necessary. Wet the spindle and inside of hub and hub cap with GREASE, general purpose, No. 2, to a maximum thickness of $\frac{1}{16}$ inch only to retard rust. Wash bearing cones and grease seals. Inspect and replace if necessary. Lubricate bearings with GREASE, general purpose, No. 2, with a packer or by hand, kneading lubricant into all spaces in the bearing. Use extreme care to protect bearings from dirt and immediately reassemble and replace wheel. The lubricant in the bearings is sufficient to provide lubrication until the next service period. Do not fill hub or hub cap. Any excess might result in leakage into the brake drum.

e. Reports and Records. If lubrication instructions are closely followed, proper lubricants used, and satisfactory results are not obtained, make a report to the ordnance officer responsible for the maintenance of the materiel. A complete record of lubrication servicing may be kept in the Duty Roster (W.D., A.G.O Form No. 6).

f. Localized Views. The localized views of lubrication points (figs. 15, 16, 17, 18, and 19) supplement the instructions on the Guide and in the notes.

SECTION VII

TOOLS AND EQUIPMENT STOWAGE ON THE VEHICLE

PARAGRAPH

19. VEHICLE TOOLS.

a. Unless the vehicle is equipped with extra tool equipment, the following are supplied (one of each unless otherwise specified):

TOOL	FEDERAL SLOCK NO.	WHERE CARRIED
HAMMER, machinist's, ball peen, 16 oz.	41-H-523	Tool bag
JACK, screw type, 1%-ton, w/handle	41-J-66	Tool compartment
PLIERS, combination, slip joint, 6-in.	41-P-1650	Tool bag
PULLER, wheel hub	41-P-2962-700	Tool compartment
WRENCH, drain plug	41-W-1962-50	Tool bag
WRENCH, engineer's open-end, 3/8- x y16-in.	41-W-991	Tool bag
WRENCH, engineer's open-end, Vz- x 19/32-in.	41-W-1003	Tool bag
WRENCH, engineer's open-end, %6- x n/ie-in.	41-W-1005-5	Tool bag
WRENCH, engineer's open-end, s/s x 25/32-in.	41-W-1008-10	Tool bag
WRENCH, engineer's open-end, 34- x y8-m.	41-W-1012-5	Tool bag
WRENCH, hydraulic brake, bleeder screw	41-W-1596-125	Tool bag
WRENCH, adjustable, auto type, 11-in.	41-W-449	Tool bag
WRENCH, socket, screw fluted	41-W-2459-500	Tool bag
WRENCH, socket, spark plug, w/handle	41-W-3335-50	Tool bag
WRENCH, wheel bearing nut, 2⅞-m. hex	41-W-3825-200	Tool compartment
WRENCH, wheel stud nut, 4%,-in. hex	41-W-3837-55	Tool compartment

20. VEHICLE EQUIPMENT.

a. Unless vehicle is equipped with special equipment, the following are supplied (one of each unless otherwise specified):

Tool	Federal Stock No.	Where Carried
ADAPTER, lubr. gun		Tool bag
APPARATUS, decontaminating, ly, qt		Driver's compartment
Ax, chopping, single-bit	41-A-1277	Body left side
BAG, tool	41-B-15	Tool compartment
CATALOG, ord. std. nom. list.	SNL-G-503	Glove compartment
CHAINS, tire, 6.00 x 16	8-C-2358	Tool compartment (4)
CONTAINER, 5-gallon		Bracket on rear
COVER, headlight		Under right seat (2)
COVER, windshield		Under right seat
CRANK, starting		Under rear seat
EXTINGUISHER, fire	58-E-202	Inside cowl, left
GAGE, tire pressure	8-G-615	Tool compartment
GUN, lubr., hand-type	41-G-1330-60	Tool compartment
MANUAL, technical	TM 9-803	Glove compartment
NOZZLE, flexible tube		
OILER, straight spout, %-pt.	13-O-1530	Front of dash
PUMP, tire, w/chuck	8-P-5000	Behind rear seat
RIFLE		On dash
SHOVEL, D-handle, rd. pt	41-S-3170	Body, left side
TAPE, friction, roll	17-T-805	Parts bag
WIRE, iron, roll	22-W-650	Parts bag

21. VEHICLE SPARE PARTS.

a. Unless the vehicle is equipped with a special assortment of parts, the following are supplied (one of each unless otherwise specified):

Name of Spare Pari	Federal Slock No.	Where Carried
BAG, spare parts	8-B-11	Glove compartment
BELT, fan	33-B-76	Parts bag
CAPS, tire valve (boxed)	8-C-650	Parts bag (5)
CORES, tire valve (boxed)	8-C-6750	Parts bag (5)

TOOLS AND EQUIPMENT STOWAGE ON THE VEHICLE

Name of Spare Part	Federal Stack No.	Where Carried
LAMP, elec. incand. 6-8V sing-tung-fil., 3 cp (MZ63)	17-L-5215	Parts bag
LAMP-UNIT, blackout, stop, sealed, one opng., 6-8V, 3 cp	8-L-421	Parts bag
LAMP-UNIT, blackout, tail, sealed, 4 opngs., 6-8V, 3 cp	8-L-415	Parts bag
LAMP-UNIT, service tail and stop, sealed, 6-8V, 21-3 cp	8-L-419	Parts bag
PIN, cotter, split, s. type B boxed ass't.	42-P-5347	Parts bag
PLUG, spark, with gasket	17-P-5365	Parts bag

PART TWO

VEHICLE MAINTENANCE INSTRUCTIONS

Section VIM

RECORD OF MODIFICATIONS

22. MWO AND MAJOR UNIT ASSEMBLY REPLACEMENT RECORD.

a. Description. Every vehicle is supplied with a copy of A.G.O. Form No. 478 which provides a means of keeping a record of each MWO completed or major unit assembly replaced. This form includes spaces for the vehicle name and U.S.A. registration number, instructions for use, and information pertinent to the work accomplished. It is very important that the form be used as directed, and that it remain with the vehicle until the vehicle is removed from service.

b. Instructions for Use. Personnel performing modifications or major unit assembly replacements must record clearly on the form a description of the work completed, and must initial the form Jn the columns provided. When each modification is completed, record the date, hours and/or mileage, and MWO number. When major unit assemblies, such as engines, transmissions, transfer cases, are replaced, record the date, hours and/or mileage, and nomenclature of the unit assembly. Minor repairs and minor parts and accessory replacements need not be recorded.

c. Early Modifications. Upon receipt by a third or fourth echelon repair facility of a vehicle for modification or repair, maintenance personnel will record the MWO numbers of modifications applied prior to the date of A.G.O. Form No. 478.

Section IX

SECOND ECHELON PREVENTIVE MAINTENANCE

23. SECOND ECHELON PREVENTIVE MAINTENANCE SERVICES.

a. Regular scheduled maintenance inspections and services are a preventive maintenance function of the using arms and are the responsibility of commanders of operating organizations.

(1) FREQUENCY. The frequency of the preventive maintenance services outlined herein is considered a minimum requirement for normal operation of vehicles. Under unusual operating conditions such as extreme temperatures, and dusty or sandy terrain, it may be necessary to perform certain maintenance services more frequently.

(2) .FIRST ECHELON PARTICIPATION. The drivers should accompany their vehicles and assist the mechanics while periodic second echelon preventive maintenance services are performed. Ordinarily the driver should present the vehicle for a scheduled preventive maintenance service in a reasonably clean condition: that is, it should be dry and not caked with mud or grease to such an extent that inspection and servicing will be seriously hampered; however, the vehicle should not be washed or wiped thoroughly clean, since certain types of defects, such as cracks, leaks, and loose or shifted parts or assemblies are more evident if the surfaces are slightly soiled or dusty.

(3) INSTRUCTIONS. If instructions other than those which are contained in the general procedures in step (4), or in the specific procedures in step (5) which follow, are required for the correct performance of a preventive maintenance service or for correction of a deficiency, other sections of the vehicle operators' manual pertaining to the item involved, or a designated individual in authority should be consulted.

(4) GENERAL PROCEDURES. These general procedures are basic instructions which are to be followed when performing the services on the items listed in the specific procedures. NOTE: *The second echelon personnel must be thoroughly trained in these procedures so that they will apply them automatically.*

(a) When new or overhauled subassemblies are installed to correct deficiencies, care should be taken to see that they are clean, correctly installed, and properly lubricated and adjusted.

(b) When installing new lubricant retainer seals, a coating of the lubricant should be wiped over the sealing surface of the lip of the seal. When the new seal is a leather seal, it should be soaked in engine oil SAE 10 (warm if practicable) for at least 30 minutes, then, the leather lip should be worked carefully by hand before installing the seal. The lip must not be scratched or marred.

(c) The general inspection of each item applies also to any supporting member or connection, and usually includes a check to see

whether or not the item is in good condition, correctly assembled, secure, or excessively worn. The mechanics must be thoroughly trained in the following explanations of these terms.

1. The inspection for "good condition" is usually an external visual inspection to determine if the unit is damaged beyond safe or serviceable limits. The term "good condition" is explained further by the following: not bent or twisted, not chafed or burned, not broken or cracked, not bare or frayed, not dented or collapsed, not torn or cut.

2. The inspection of a unit to see that it is "correctly assembled" is usually an external visual inspection to see if it is in its normal assembled position in the vehicle.

3. The inspection of a unit to determine if it is "secure" is usually an external visual examination, a hand-feel, wrench, or a pry-bar check for looseness. Such an inspection should include any brackets, lock washers, lock nuts, locking wires, or cotter pins used in assembly.

4. "Excessively worn" will be understood to mean worn, close to or beyond serviceable limits, and likely to result in a failure if not replaced before the next scheduled inspection.

(d) Special Services. These are indicated by repeating the item numbers in the columns which show the interval at which the services are to be performed, and show that the parts or assemblies are to receive certain mandatory services. For example, an item number in one or both columns opposite a *Tighten* procedure means that the actual tightening of the object must be performed. The special services include:

1. Adjust. Make all necessary adjustments in accordance with the pertinent section of the vehicle operator's manual, special bulletins, or other current directives.

2. Clean. Clean units of the vehicle with dry-cleaning solvent to remove excess lubricant, dirt, and other foreign material. After the parts are cleaned, rinse them in clean fluid and dry them thoroughly. Take care to keep the parts clean until reassembled, and be certain to keep cleaning fluid away from rubber or other material which it will damage. Clean the protective grease coating from new parts, since this material is not a good lubricant.

3. Special lubrication. This applies both to lubrication operations that do not appear on the vehicle Lubrication Guide and to items that do appear on such charts, but which should be performed in connection with the maintenance operations if parts have to be disassembled for inspection or service.

4. Serve. This usually consists of performing special operations, such as replenishing battery water, draining and refilling units with oil, and changing the oil filter cartridge.

5. Tighten. All tightening operations should be performed with sufficient wrench-torque (force on the wrench handle) to tighten the unit according to good mechanical practice. Use torque-indicating wrench where specified. Do not overtighten, as this may strip threads or cause distortion. Tightening will always be understood to include the correct installation of lock washers, lock nuts, and cotter pins provided to secure the tightening.

(e) *Conditions.* When conditions make it difficult to perform the complete preventive maintenance procedures at one time, they can sometimes be handled in sections, planning to complete all operations within the week, if possible. All available time at halts and in bivouac areas must be utilized, if necessary, to assure that maintenance operations are completed. When limited by the tactical situation, items with special services in the columns should be given first consideration.

(f) The numbers of the preventive maintenance procedures that follow are identical with those outlined on W.D., A.G.O. Form No. 461, which is the Preventive Maintenance Service Work Sheet for Wheeled and Half-track Vehicles. Certain items on the work sheet that do not apply to this vehicle are not included in the procedures in this manual. In general, the numerical sequence of items on the work sheet is followed in the manual procedures, but in some instances there is deviation for conservation of the mechanic's time and effort.

(5) SPECIFIC PROCEDURES. The procedures for performing each item in the 1,000-mile (monthly) and 6,000-mile (6-month) maintenance procedures are described in the following chart. Each page of the chart has two columns at the left edge corresponding to the 6,000-mile and the 1,000-mile maintenance respectively. Very often it will be found that a particular procedure does not apply to both scheduled maintenances. In order to determine which procedure to follow, look down the column corresponding to the maintenance due, and wherever an item appears, perform the operations indicated opposite the number.

ROAD TEST

MAINTENANCE		
6000 Mile	1000 Mile	
		NOTE: *When the tactical situation does not permit a full road test, perform those items which require little or no movement of the vehicle, namely, items 3, 4, 5, 6, 9, 10, and 14. Make a full road test of 5, but not more than 10 miles, over varied terrain if possible.*
1	1	Before-operation Service. Perform Before-operation Service as outlined in paragraph 13.
3	3	Dash Instruments and Gages. Observe instruments frequently during road test.
		AMMETER. Ammeter should show high charge for short time after starting, then zero or slight positive (plus) reading above speeds of 12 to 15 miles per hour with lights and accessories off. Zero reading is normal with lights and accessories on.
		SPEEDOMETER. See that speedometer indicates vehicle speed, operates without excessive fluctuation or noise, and that odometer registers accumulating trip and total mileage correctly.

¼-TON 4 x 4 TRUCK (WILLYS-OVERLAND MODEL MB and FORD MODEL GPW)

MAINTENANCE		
6000 Mile	1000 Mile	
		TEMPERATURE INDICATOR. Temperature indicator should gradually increase to normal operating range of 160°F to 180°F.
		FUEL GAGE. Fuel gage must indicate the approximate amount of fuel in tank.
4	4	Horn, Mirror, and Windshield Wiper. Test horn for proper operation and tone, tactical situation permitting. Adjust mirror, and inspect for broken or discolored glass. Wiper should have sufficient arm tension to stay in "UP" position. Examine blade for good condition and full contact with glass throughout entire stroke.
5	5	Brakes. Test brakes for smooth, even stop, excessive pedal travel before application, "spongy" pedal, or loss of pedal pressure when brakes are held on. Brakes must not squeak or require excessive pedal pressure. Test pedal free travel, which should be % inch. Hand brake must hold vehicle on a reasonable grade, must have positive ratchet action and V3 reserve handle travel. There should be %-inch reserve clearance between hand brake relay crank and lower end of hand brake cable conduit.
6	6	Clutch. Clutch must have free pedal travel of three-quarter inch. Test clutch for slip, grab, gear clash, or rattle. Listen for noises that would indicate dry or defective release bearing or pilot bushing.
7	7	Transmission and Transfer Case. Shift through entire range of transmission and transfer, noting whether the levers move easily and snap into each position. With shifting levers in each position, accelerate and decelerate engine, noting any unusual noises or tendency of levers to slip into neutral. Inspect for loose mountings.
8	8	Steering. Steering gear must not bind. There should be no excessive free play with wheels in straightahead position. Test for existence of front-end shimmy, wander, or side pull.
9	9	Engine. Engine must idle smoothly without stalling. Test acceleration and pulling power in each transmission speed. Listen for detonation and "ping," misses, popping, spitting, or other noises that might indicate need for engine repair.
10	10	Unusual Noises. Listen for noises that might indicate loose, damaged, or faulty parts.
13	13	Temperatures. Feel brake drums and wheel hubs for abnormally high temperatures. Overheated brake drum or wheel hub may indicate dragging brake or defective, dry, or improperly adjusted wheel bearing. Examine

MAINTENANCE		
6000 Mile	1D00 Mile	
		differentials, transmission, and transfer case for too-high running temperature. NOTE: *Transfer case operates at a higher temperature than other* cases.
14	14	Leaks. Look on ground under vehicle for indications of coolant, fuel, oil, or hydraulic fluid leaks.
16	16	Gear Oil Level and Leaks. Examine lubricant levels of transmission, transfer case, and differentials. Inspect cases for leaks. Safe level when cold is even with filler plug. If an oil change is due, drain and refill, according to Lubrication Guide (par. 18). Capacities: transmission, *3/4* quart; transfer case, *I/,* quarts; front differential, 1% quarts; rear differential, 1% quarts.
		MAINTENANCE OPERATIONS
17	17	Unusual Noises. With engine running, proceed as follows: Accelerate and decelerate engine slightly, and listen for unusual engine noises. With transmission in third gear, front wheel drive engaged, and engine at fast idle, listen for unusual noises in operating units. Observe propeller shaft and universal joints, wheels, and axles for excessive vibration and run-out.
22	22	Battery. Inspect battery case for cracks and leaks. Inspect cables, terminals, bolts, posts, straps, and hold-downs for good condition and secure mounting. Clean top of battery. Test specific gravity and voltage, and record on W.D., A.G.O. Form No. 461. Specific gravity readings below 1.225 indicate battery should be recharged or replaced. Electrolyte level should be above top of plates, and may extend *y,* inch above plates.
22	22	SERVE. Perform high-rate discharge test according to instructions for "condition" test which accompany test instrument, and record voltage on W.D., A.G.O. Form No. 461. Cell variation should not be more than 30 percent. NOTE: *Specific gravity must be above 1.225 to make this test.*
		CLEAN. Clean entire battery and carrier, and repaint carrier if corroded. Clean battery cable terminals, terminal bolts and nuts, and battery posts; grease lightly; inspect bolts for serviceability. Tighten terminals and hold-downs carefully to avoid damage to battery. Add clean water to *y,* inch above plates.
18	18	Cylinder Head and Gasket. Look for cracks, and indications of water or compression leaks. Tighten cylinder head (only if leaks are indicated and after performing item 21) with torque wrench; tighten head-screws to from 65 to 75 foot-pounds; head stud nuts

MAINTENANCE 6000 Mile	1000 Mile	
		to from 60 to 65 foot-pounds. Tighten in correct order (fig. 25). Be sure cylinder head to dash bond strap is in good condition and securely connected.
	19	Valve Mechanism. Adjust valves only if noisy.
19		ADJUST. Check clearance and adjust valves. Proper clearances are: intake valve, 0.014 inch when hot or cold; exhaust valve, 0.014 inch when hot or cold.
	20	Spark Plugs. Wipe off plugs without removing; inspect for insulator cracks and leakage through insulators and gaskets. Service if required.
20		SERVE. Clean and adjust plugs to gap of 0.030 inch, using round gage. Plugs with broken insulators, excessive carbon deposits, electrodes burned thin or otherwise unserviceable, must be replaced. Correct plug (AN-7). NOTE: *// sand blast cleaner is not available install new or reconditioned plugs.*
21	21	Compression. Test compression with all plugs removed, and with throttle and choke wide open. Standard pressure is approximately 110 pounds at cranking speed; minimum pressure is 70 pounds. Maximum variation between cylinders must not be more than 10 pounds. If variation is greater than 10 pounds, recheck weak cylinders, using oil test, and report to higher authority. Record all readings.
23	23	Crankcase. Observe vehicle for crankcase, valve cover, timing case, or flywheel housing oil leaks. Check oil level. Drain and refill crankcase if change is due. See Lubrication Guide (par. 18).
23		CAUTION: *Do not start engine until completion of item 24.*
24	24	Oil Filters and Lines. Inspect filters, lines, and connections for good condition or leaks.
24		SERVE: Remove filter cartridge, clean filter case and install new cartridge and gaskets. Refill crankcase (5 quarts with new filter cartridge). Again inspect for leaks with engine running and check oil level after engine is stopped.
25	25	Radiator. Observe radiator core, hose, cap and gaskets for good condition and inspect for leaks. CAUTION: *System operates under 3]/^ to 4% pounds pressure (be careful in removing cap).* Examine air passages and guards for obstructions and clean out any dirt, insects, or trash. Test and record antifreeze value (as climate demands). Examine coolant for oiL rust, or foreign

SECOND ECHELON PREVENTIVE MAINTENANCE

MAINTENANCE		
6009 Mill	1000 Mile	
		material. Clean and flush radiator as needed. CAUTION: *Save and filter coolant if antifreeze is present.* Add inhibitor and antifreeze if needed.
25		TIGHTEN. Tighten hose clamps. Inspect radiator cap and gasket for tight seal.
26	26	Water Pump and Fan. Loosen fan belt; test water pump shaft and bearing for play. Inspect pump for secure attachment, good condition, and for leaks. Inspect fan for alinement and secure mounting.
27	27	Generator, Cranking Motor, and Switch. Inspect these units to see if they are in good condition, clean and securely connected or mounted; particularly radio noise suppression capacitor on generator and starting switch terminal, and bond straps from generator and cranking motor.
27		SERVE. Inspect commutators and brushes for good condition and wear. Brushes should be free in holders, and have full contact with commutator. Clean commutators with 2/0 flint paper if needed. Blow out with compressed air. Replace generator or cranking motor when commutator is scored, rough, worn, or brushes are less than half their original length.
29	29	Drive Belt and Pulleys. Inspect fan belt for fraying, wear, and deterioration. Inspect pulleys for cracks and misalinement. Replace or adjust belt as needed. Adjust to deflection of 1 inch between pulleys.
31	31	Distributor. Clean and remove distributor cap. Examine cap and rotor arm for cracks, corrosion and burned conductors. Clean breaker plate assembly, if dirty. Inspect breaker points for burning, pitting, alinement, and adjustment. Replace and aline burned or badly pitted points. Feel to determine excessive distributor shaft play. Turn distributor shaft (with rotor), and release to test centrifugal advance for binding.
31		SPECIAL LUBRICATION. Sparingly lubricate cam surfaces, movable breaker arm pin, wick and camshaft according to Lubrication Guide (par. 18). Adjust breaker point gap to 0.020 inch.
32	32	Coil and Wiring. Examine coil, high tension, and exposed low voltage wiring for cleanliness, and secure connections and attachment. Clean and tighten as required. Pay particular attention to see that spark plug

MAINTENANCE 6000 Mile	MAINTENANCE 1000 Mile	
		and coil to distributor wire, radio noise suppressors, and coil terminal capacitor are in good condition, and securely mounted or connected.
33	33	Manifolds and Heat Control. Tighten manifold stud nuts as required to from 31 to 35 foot-pounds. Inspect for gasket leaks. Heat control valve must be free and bimetal spring must be in good condition.
34	34	Air Cleaner. Examine air cleaner for good condition and secure mounting. Examine oil cup. If dirty, remove and clean filter element; do not apply oil to element after cleaning. Clean oil cup and refill *(s/g* qt).
36	36	Carburetor. Make certain that the choke and throttle open and close fully. Lubricate linkage, and inspect for worn parts.
37	37	Fuel Filter, Screens, and Lines. Clean fuel pump screen, renew gaskets, inspect unit for leaks. Remove disk filter element from fuel filter mounted on dash; clean element and bowl. Reinstall with new gasket. Inspect for leaks after unit has been refilled.
38	38	Fuel Pump. Observe fuel pump for leaks, secure mounting, and pressure reading. Pressure should be 1% to 2% pounds with engine running at approximately 30 miles per hour vehicle road speed.
39	39	Cranking Motor. Start engine and observe cranking motor for positive action, normal speed, and unusual noise. Make sure oil pressure gage and ammeter readings are satisfactory.
40	40	Leaks. Look around engine and on ground under engine for oil, fuel, coolant, or hydraulic fluid leaks.
41	41	Ignition Timing. With neon light, check ignition timing. Observe if spark advances automatically. Adjust timing as required (par. 65). CAUTION: *Close timing hole cover and tighten screw.*
42	42	Engine Idle and Vacuum Test. Adjust engine to smooth idle, using vacuum gage; obtain highest possible steady vacuum reading.
	43	Regulator Unit. See that regulator and radio noise capacitors are in good condition, and that all connections and mounting are secure.
43		TEST. Connect low voltage circuit tester and test voltage regulator, current regulator, and cut-out for output control.
47	47	Tires and Rims. Inspect valve stems for correct position and missing caps. Inspect tires for cuts, bruises, blisters, irregular and excessive tread wear. Remove imbedded glass, nails, or stones. Directional and non-

SECOND ECHELON PREVENTIVE MAINTENANCE

MAINTENANCE 6000 Mile	1000 Mile	
		directional tires should not be installed on same vehicle. If equipped with directional tires, open end of chevron should meet ground first on front tires, and last on rear tires. Tires should match on all wheels within 3^-inch over-all circumference, and as to type of tread. Take measurements with all tires equally inflated. Inspect tire carrier for looseness and damage. Tighten all lug nuts securely. Inflate tires to 35 pounds (cold).
48	48	Rear Brakes. Remove grease and dirt from brake drums and backing plates, and inspect for excessive wear or scoring and loose mounting bolts. Inspect brake hose for proper fit and for deterioration. Inspect wheel cylinders (exterior) for good condition, secure mounting, and for leaks. Tighten brake support and drum mounting bolts securely.
49	49	Rear Brake Shoes. Remove right rear wheel and inspect linings for wear, oil, and dirt, and possibility of rivets scoring drum before next 1,000-mile inspection. If lining on right rear wheel requires replacement, remove all wheels for lining inspection.
49		SERVE. Remove all wheels and drums. Observe linings for wear, oil, and dirt, and determine if shoes are secure and guided by anchor pins. Inspect return springs for good action. Lightly lubricate anchor pins. Adjust brake shoes to 0.005 inch at heel, and 0.008 inch at toe.
	52	Rear Wheels. Inspect wheel for good condition and, without removal, test for evidence of looseness of wheel bearing adjustment, and dry or damaged bearings. Inspect around drive flanges, brake supports, and drums for lubricant or brake fluid leaks. Tighten drive flange and wheel nuts. CAUTION: *// it is known that vehicle has operated in deep water which may have entered wheel bearings, inspect right wheel bearing fpr contamination. Remove, clean, repack, and adjust as for 6,000-mile service. If contamination of lubricant has occurred, service other wheel bearings likewise.*
52		CLEAN. Disassemble wheel bearings and seals, clean, and inspect for damage.
		SPECIAL LUBRICATION. Pack wheel bearings, install new seals, and adjust bearings.
53	53	Front Brakes. Examine brake hose for chafing, leakage, and deterioration. Inspect wheel cylinders (exterior) for good condition, secure mounting, and leaks.
53		DRUMS AND SUPPORTS. Clean drums and backing plates thoroughly, and tighten backing plate bolts. Inspect drums for damage, looseness, excessive wear, and scoring. Lightly lubricate anchor pins.
54		Front Brake Shoes. Inspect brake shoes, linings, and anchors for damage or looseness. Replace worn parts

MAINTENANCE		
6000 Mile	1000 Mile	
		and worn linings. Clean dust from linings. Adjust brake shoes to 0.005-inch clearance at heel, and 0.008-inch clearance at toe.
55	55	Steering Knuckles. Inspect steering knuckle housings and oil seals for serviceable condition. Check lubricant for contamination. Refill to bottom of filler hole.
56	56	Front Springs. Inspect front springs for good condition, correct alinement, and excessive deflection. Inspect springs for excessive wear of spring bushing and clips. Tighten U-bolts securely and Uniformly. Examine U-shackles and pivot bolts for wear.
57	57	Steering. Observe steering gear, Pitman arm, drag link, tie rod, and steering connecting rods for good condition, correct assembly, and secure mounting.
57		TIGHTEN. Tighten and adjust assembly mounting nuts and screws, arms, tie rods, drag link, Pitman arm, and gear, and steering wheel nuts. Replace broken seals or worn parts.
58	58	Front Shock Absorbers. Inspect shock absorbers to see if they are in good condition and secure, if bodies are leaking fluid, and if rubber bushings have deteriorated. If rubber bushings are hard or cracked, apply a film of brake fluid. NOTE: *// fluid is leaking or bodies are defective, shock absorber must be replaced.*
60	60	Front Wheels. Inspect for good condition, security, end play, and lubricant leaks. Rotate wheels and observe for loose, broken, or dry bearings.
60		CLEAN AND LUBRICATE. Remove, clean, inspect, lubricate, and replace bearings. Adjust bearings and test for wheel shake before removing jack.
61	61	Front Axle. Examine front axle housing for good condition and lubricant leaks. Inspect pinion shaft for end play and grease leaks. Inspect axle for apparent alinement, and see that vent is open.
62	62	Front Propeller Shaft. Inspect propeller shaft for damage and incorrect assembly, excessive wear, and lubricant leaks. Inspect universal and slip joints for alinement, wear, and leakage.
62		TIGHTEN. Tighten flange yoke bolts.
63	63	Engine Mountings and Braces. See that engine mountings and bond straps are in good condition and secure, and that rubber mountings are not separated from metal backing. Tighten front mountings if loose. Adjust rear

MAINTENANCE		
6000 Mile	1000 Mile	
		mounting bolts to from 38 to 42 foot-pounds with torque wrench. Tighten radio noise suppression bond strap mountings securely.
64	64	Parking (Hand) Brake. See that drum is not scored or oily; that lining is not oil-soaked nor worn thin. Inspect ratchet for positive holding action. Lubricate upper end of conduit tube at cable with engine oil.
64		ADJUST. Adjust clearance between drum and lining to from 0.005 inch to 0.010 inch. Reserve lever travel should be one-third the ratchet range. There must be 5 4 - i " " reserve clearance (on cable) between relay crank and lower end of hand brake conduit.
65	65	Clutch Pedal. Clutch pedal linkage must be secure and not worn; return spring must be operative; clutch should have free pedal travel of *y$* inch.
65		ADJUST. Adjust clutch pedal free travel to % inch.
66	66	Brake Pedal. Test brake pedal operation; brake linkage must be secure and not worn excessively; return spring must be operative; brake should have reserve travel.
66		ADJUST. Adjust brake pedal free travel to *y,* inch.
67	67	Brake Master Cylinder. Inspect master cylinder for good condition and secure mounting; check master cylinder boot for good condition and correct installation; inspect stop light switch for terminal attachment and correct operation. Look for brakeFLuidleaks; clean outFIller plug vent. Fill master cylinder reservoir to % inch below plug.
71	71	Transmission. Inspect oil seals and gaskets for leakage. Test control for looseness, excessive wear, and improper operation. Inspect mounting and assembly bolts and cap screws for looseness.
71		TIGHTEN. Tighten mounting and assembly bolts and cap screws.
72	72	Transfer Case. Inspect oil seals and gaskets for leakage. Test controls for looseness, excessive wear, and improper operation. Inspect mounting and assembly bolts and cap screws for looseness. Clean vent.
72		TIGHTEN. Tighten mounting and assembly bolts, nuts, and cap screws.
73	73	Rear Propeller Shaft. Remove any trash that may be wrapped around shaft or universal joints. Inspect

MAINTENANCE 6000 Mile	MAINTENANCE 1000 Mile	
		mounting of universal and slip joints for misalinement, wear, and grease leaks.
73		TIGHTEN. Tighten flange yoke cap screws.
75	75	Rear Axle. Inspect rear axle housing for leaks; feel for excessive play in pinion shaft; clean vent. Make sure differential carrier mounting cap screws are tight.
77	77	Rear Springs. Check springs for shifted leaves due to broken center bolt, loose spring clips, or U-bolts. If found loose, tighten U-bolts to from 50 to 55 foot-pounds. Tighten spring pivot bolt nut to from 29 to 30 foot-pounds.
78	78	Rear Shock Absorbers. Inspect in the same manner as for item 58.
80	80	Frame. Examine frame for loose side rails and cross members. Tighten loose bolts. If frame appears to be bent, or out of alinement, report condition to higher authority.
81	81	Wiring, Conduits and Grommets. Inspect all wiring for looseness and broken insulation; check conduits and grommets for proper position and good condition.
82	82	Fuel Tank and Lines. Inspect tank and lines for good condition, secure mounting, and leaks; check cap for defective gasket or clogged vent.
82		SERVE. Remove fuel tank drain plug briefly, and drain off accumulated water and sediment.
83	83	Brake Lines and Connections. Inspect brake lines for proper mounting, cracks, worn spots in lines, leaks, deteriorated or damaged hose and connections.
84	84	Exhaust Pipe and Muffler. Inspect exhaust pipe and muffler for secure mounting, rusted condition, damage or leaks. Inspect tail pipe for stoppage.
85	85	Vehicle Lubrication. Lubricate according to Lubrication Guide (par. 18) in this manual. Observe latest issued lubrication directives.
		LOWER VEHICLE TO GROUND
86	86	Toe-in and Turning Stops. With front wheels on ground, straight-ahead position, use wheel alining gage, and check toe-in. Normal toe-in range is %,-inch to %,,-i"h- Turn front wheels fully in both right and left

MAINTENANCE		
MM Mlk	ma Mill	
		directions, and determine if turning stops hold tires clear of all parts of vehicle in these positions. Examine axle for loose turn stops.
91	91	Lights. Determine that switches for head, tail, instrument, and blackout lights operate properly. Operate stop light by depressing brake pedal. Test foot switch, noting whether beam is controlled for high and low positions. Inspect all lights; these must be clean, securely mounted, and in good condition; lenses must not be broken, cracked, or discolored; reflectors must not be discolored; blackout lights must be in good condition with shield in proper position.
91		ADJUST. Adjust and aim headlight beams.
92	92	Safety Reflectors. Safety reflectors must be present, clean, and secure. Replace if cracked or broken.
93	93	Front Bumper and Grille. Front bumper and grille must be present, in good condition, and securely mounted.
94	94	Hood, Hinges and Fasteners. Examine hood for alinement and secure mounting when fastened; see that fasteners are present, secure, undamaged, and not excessively worn or bent. Lubricate hinges and fasteners lightly. See that radio noise bond straps from hood to dash and grille are secure.
95	95	Front Fenders. Inspect front fenders for good condition and secure mounting.
96	96	Body Hardware. Inspect body of vehicle according to following standards: Hardware should operate properly and be adequately lubricated; top should be clean, having no holes or tears, and all grommets must be present and in good condition. Windshield should be free from cracks or discoloration; windshield frame and hold-down hooks at hood should be in good condition. Seats and upholstery should be clean and undamaged; safety straps should be present and in place; body handles should be present, secure, and undamaged; floor drain plugs (2) should be present, and in good condition.
98	98	Circuit Breaker, Terminal Blocks, or Boxes. Inspect points of thermal circuit breaker (30 amperes, located on main light switch) for pitting or corrosion. Be sure all radio noise suppression bond straps and capacitor on radio terminal box (if so equipped) are in good condition and secure.
101	101	Rear Bumpers and Pintle Hook. Latch and Lock Pin. Inspect rear bumpers and pintle hook to see if they are

MAINTENANCE		
6000 Mile	1090 Mile	
		present, in good condition, and secure. Pintle hook safety latch should be free, and lock securely.
103	103	Paint and Markings. Inspect paint of entire vehicle for good condition and bright spots that might cause glare or reflection. Vehicle markings and identification must be legible. Inspect identification plates and their mountings (if furnished) for good condition, secure mounting, and legibility.
104	104	Radio Bonding (Suppressors, Filters, Condensers, and Shielding). See that all units not covered in the foregoing specific procedures are in good condition, and securely mounted and connected. Be sure all additional noise suppression bond straps and toothed lock washers listed in paragraph 177, are inspected for looseness or damage, and see that contact surfaces are clean. NOTE: *// objectionable radio noise from vehicle has been reported, make tests in accordance with paragraph 178. If cleaning and tightening of mountings and connections, and replacement of defective radio noise suppression units does not eliminate the trouble, the radio operator will report the condition to the designated individual in authority.*
105	105	Armament. Examine gun mounts and covers (if present) for good condition, cleanliness, and secure attachment. NOTE: Guns, *parts, and covers are to be referred to armorer or gun commanders for all inspections or service.*
		TOOLS AND EQUIPMENT
131	131	Tools and Equipment. Standard vehicle tools, Pioneer tools, and equipment must be present, clean, serviceable, and securely mounted. Sharpen cutting tools and darken bright parts of exposed tools in combat areas. Check against stowage list (par. 19).
132	132	Fire Extinguisher. Inspect fire extinguisher for full charge and secure mounting. See that nozzle is clean.
133	133	Decontaminator. Inspect decontaminator for damage, secure mounting, and full charge. Make latter check by removing filler plug. Drain and refill with fresh solution every 90 days. See date of last filling on attached tag.
143	143	First Aid Kit. Examine contents of first aid kit for good condition, completeness, and satisfactory packing. Report any deficiency.
135	135	Publications and Form No. 26. See that the vehicle manuals and Lubrication Guide, Form No. 26 (Acci-

MAINTENANCE		
6000 Mile	1000 Mile	
		dent Report) and W.D., A.G.O. Form No. 478 (MWO and Major Unit Assembly Replacement Record), are present, legible, and properly stowed.
136	136	Traction Devices. Inspect tire chains for broken or worn links, missing cross chains, or damaged fasteners.
139	139	Fuel Can and Bracket. Inspect fuel can and bracket for damage, leaks, loose mounting, and presence of cap on chain.
140	140	Fuel Can Nozzle and Bucket. See that fuel can nozzle and bucket are not damaged, are clean, and properly stowed.
141	141	Modifications (Completed). Inspect entire vehicle to be sure all Modification Work Orders have been completed, and enter any modifications or major unit replacements made at time of this service, on Form No. 478.
142	142	Final Road Test. Road test, rechecking items 2 to 16. Recheck transmission, transfer case, and differentials for lubricant level and for leaks. Confine this test to minimum distance necessary to satisfactory observations. NOTE: *Correct or report all defects found during final road test to higher authority.*

Section X

NEW VEHICLE RUN-IN TEST

24. PURPOSE.

a. When a new or reconditioned vehicle is first received at the using organization, it is necessary for second echelon personnel to determine whether or not the vehicle will operate satisfactorily when placed in service. For this purpose, inspect all accessories, subassemblies, assemblies, tools, and equipment to see that they are in place and correctly adjusted. In addition, they will perform a run-in test of at least 50 miles as directed in AR 850-15, paragraph 25, table III, according to procedures in paragraph 26 below.

25. CORRECTION OF DEFICIENCIES.

a. Deficiencies disclosed during the course of the run-in test will be treated as follows:

(1) Correct any deficiencies within the scope of the maintenance echelon of the using organization before the vehicle is placed in service.

(2) Refer deficiencies beyond the scope of the maintenance echelon of the using organization to a higher echelon for correction.

(3) Bring deficiencies of serious nature to the attention of the supplying organization.

26. RUN-IN TEST PROCEDURES.

a. Preliminary Service.

(1) FIRE EXTINGUISHER. See that portable extinguisher is present and in good condition. Test it momentarily for proper operation, and mount it securely.

(2) FUEL, OIL, AND WATER. Fill fuel tank. Check crankcase oil and coolant supply; add oil and coolant as necessary to bring to correct levels. Allow room for expansion in fuel tank and radiator. During freezing weather, test value of antifreeze, and add as necessary to protect cooling system against freezing. CAUTION: 77 *there is a tag attached to filler cap or steering wheel concerning engine oil in crankcase, follow instructions on tag before driving the vehicle.*

(3) FUEL FILTER. Inspect main fuel filter for leaks, damage, and secure mountings and connections. Drain sediment bowl. Clean fuel pump filter screen and bowl. If any appreciable amount of dirt or water is present, remove main filter bowl and clean bowl and element

in dry-cleaning solvent. Also, drain accumulated dirt and water from bottom of fuel tank. Drain only until fuel runs clean.

(4) BATTERY. Make hydrometer and voltage test of battery, and add clean water to bring electrolyte *Y%* inch above plate.

(5) AIR CLEANER. Examine carburetor air cleaner to see if it is in good condition and secure. Remove element and wash thoroughly in dry-cleaning solvent. Fill oil cup to indicated level with fresh oil, and reinstall securely. Be sure oil cup and body gaskets are in good condition, and that air horn connection is tight.

(6) ACCESSORIES AND BELT. See that accessories such as carburetor, generator, regulator, cranking motor, distributor, water pump, fan, and oil filter, are securely mounted. Make sure that fan find generator drive belt is in good condition, and adjusted to have 1-inch finger-pressure deflection.

(7) ELECTRICAL WIRING. Examine all accessible wiring and conduits to see if they are in good condition, securely connected, and properly supported.

(8) TIRES. See that all tires, including spare, are properly inflated to 35 pounds, cool; that stems are in correct position; all valve caps present and finger-tight. Inspect for damage, and remove objects lodged in treads and carcasses.

(9) WHEEL AND FLANGE NUTS. See that all wheel mounting and axle flange nuts are present and secure.

(10) FENDERS AND BUMPER. Examine fenders and front bumper for looseness and damage.

(11) TOWING CONNECTIONS. Examine towing shackles and pintle hook for looseness and damage, and see that pintle latch operates properly and locks securely.

(12) BODY. See that all body mountings are secure. Inspect attachments, hardware, glass, seats, grab rails and safety straps, top and frame, curtains and hood, to see if they are in good condition, correctly assembled, and securely mounted or fastened. Examine body paint or camouflage pattern for rust, or shiny surfaces that might cause glare. See that vehicle markings are legible.

(13) LUBRICATE. Perform a complete lubrication service of the vehicle, covering all intervals, according to instructions on Lubrication Guide (par. 18), except gear cases, wheel bearings, and other units already lubricated or serviced in items (1) to (12). Check all gear case oil levels, and add as necessary to bring to proper levels. Change only if condition of oil indicates the necessity, or if gear oil is not of proper grade for existing atmospheric temperatures. NOTE: *Perform following items* (14) *through* (17) *during lubrication.*

(14) SPRINGS AND SUSPENSIONS. Inspect front and rear springs and shocks to see that they are in good condition, correctly assembled, secure, and that bushings and shackle pins are not excessively loose, or damaged.

(15) STEERING LINKAGE. See that all steering arms, rods, and connections are in good condition and secure; and that gear case is securely mounted and not leaking excessively.

(16) PROPELLER SHAFTS. Inspect all shafts and universal joints to see if they are in good condition, correctly assembled, alined, secure, and not leaking excessively.

(17) AXLE AND TRANSFER VENTS. See that axle housing and transfer case vents are present, in good condition, and not clogged.

(18) CHOKE. Examine choke to be sure it opens and closes fully in response to operation of choke button.

(19) ENGINE WARM-UP. Start engine and note if cranking motor action is satisfactory, and if engine has any tendency toward hard starting. Set hand throttle to run engine at fast idle during warm-up. During warm-up, reset choke button so that engine will run smoothly, and to prevent overchoking and oil dilution.

(20) INSTRUMENTS.

(a) Oil Pressure Gage. Immediately after engine starts, observe if oil pressure is satisfactory. (Normal operating pressure, hot, at running speeds is 40 to 50 pounds; at idle, 10 pounds). Stop engine if pressure is not indicated in 30 seconds.

(b) Ammeter. Ammeter should show slight positive (+) charge. High charge may be indicated until generator restores to battery, current used in starting.

(c) Temperature Gage. Engine temperature should rise gradually during warm-up period to normal operating range, 160°F to 185°F

(d) Fuel Gage. Fuel gage should register "FULL" if tank has been filled.

(21) ENGINE CONTROLS. Observe if engine responds properly to controls, and if controls operate without excessive looseness or binding.

(22) HORN AND WINDSHIELD WIPERS. See that these items are in good condition and secure. If tactical situation permits, test horn for proper operation and tone. See if wiper arms will operate through their full range, and that blade contacts glass evenly and firmly.

(23) GLASS AND REAR VIEW MIRROR. Clean all body glass, curtain windows, and mirror, and inspect for looseness and damage. Adjust mirror for correct vision.

(24) LAMPS (LIGHTS) AND REFLECTORS. Clean lenses and inspect all units for looseness and damage. If tactical situation permits, open and close all light switches to see if lamps respond properly.

(25) LEAKS, GENERAL. Look under vehicle, and within engine compartment, for indications of fuel, oil, coolant, and brake fluid leaks. Trace to source any leaks found, and correct or report them to designated authority.

(26) TOOLS AND EQUIPMENT. Check tools and On Vehicle Stowage Lists, paragraphs 19 and 20, to be sure all items are present, and see that they are serviceable, and properly mounted or stowed.

b. Run-in Test. Perform the following procedures, steps (1) to (11) inclusive, during the road test of the vehicle. On vehicles which have been driven 50 miles or more in the course of delivery from the supplying to the using organization, reduce the length of the road test to the least mileage necessary to make observations listed below. CAUTION: *Continuous operation of the vehicle at speeds approaching the maximum indicated on the caution plate should be avoided during the test.*

(1) DASH INSTRUMENTS AND GAGES. Do not move vehicle until engine temperature reaches 135°F. Maximum safe operating temperature is 200°F. Observe readings of ammeter, oil temperature, and fuel gages to be sure they are indicating the proper function of the units to which they apply. Also see that speedometer registers the vehicle speed, and that odometer registers accumulating mileage.

(2) BRAKES: FOOT AND HAND. Test service brakes to see if they stop vehicle effectively, without side pull, chatter, or squealing; and observe if pedal has at least y,-inch free travel before meeting push rod-to-piston resistance. Parking brake should hold vehicle on reasonable incline, leaving one-third lever ratchet travel in reverse. CAUTION: *Avoid long application of brakes until shoes become evenly seated to drums.*

(3) CLUTCH. Observe if clutch operates smoothly without grab, chatter, or squeal on engagement, or slippage (under load) when fully engaged. See that pedal has ^-inch free travel before meeting resistance. CAUTION: *Do not ride clutch pedal at any time, and do not engage and disengage new clutch severely or unnecessarily.*

(4) TRANSMISSION AND TRANSFER. Gearshift mechanism should operate easily and smoothly, and gears should operate without excessive noise, and not slip out of mesh. Test front axle declutching for proper operation.

(5) STEERING. Observe steering action for binding or looseness, and note any excessive pull to one side, wander, shimmy, or wheel tramp. See that column, bracket, and wheel are secure.

(6) ENGINE. Be on the alert for any abnormal engine operating characteristics or unusual noise, such as lack of pulling power or acceleration, backfiring, misfiring, stalling, overheating, or excessive exhaust smoke. Observe if engine responds properly to all controls.

(7) UNUSUAL NOISE. Be on the alert throughout road test for any unusual noise from body and attachments, running gear, suspension, or wheels, that might indicate looseness, damage, wear, inadequate lubrication, or underinflated tires.

(8) HALT VEHICLE AT 10-MILE INTERVALS FOR SERVICES (steps (9) and (10) below).

(9) TEMPERATURES. Cautiously hand-feel each brake drum and wheel hub for abnormal temperatures. Examine the transmission, transfer case, and differential housing for indications of overheating

and excessive lubricant leaks at seals, gaskets, or vents. NOTE: *Transfer case temperatures are normally higher than other gear* cases.

(10) LEAKS. With engine running, and fuel, engine oil, and cooling systems under pressure, look within engine compartment and under vehicle for indications of leaks.

c. Upon completion of run-in test, correct or report any deficiencies noted. Report general condition of vehicle to designated individual in authority.

Section XI

ORGANIZATION TOOLS AND EQUIPMENT

27. STANDARD TOOLS AND EQUIPMENT.

a. All standard tools and equipment available to second echelon are listed in SNL N-19, and their availability is determined by the table of equipment for any particular organization.

28. SPECIAL TOOLS.

a. The special tools available to second echelon for repair of this vehicle are listed in the Organizational Spare Parts and Equipment List of SNL G-503. The special tools required for the operations described in this manual are listed below:

Tool	Federal Stock No.
COMPRESSOR, shock absorber grommet	41-C-2554-400
WRENCH, tappet, double-end, i'^- x i%,-in.	41-W-3575

Section XII

TROUBLE SHOOTING

29. GENERAL.

a. The following listed possible vehicle troubles and remedies will assist in determining the cause of unsatisfactory operation. A separate list is provided for each unit. If the remedy is not given, reference is made to a paragraph where more complete information will be found.

b. The information in this section applies to operation of the vehicle under normal conditions. If extreme conditions are encountered, it is assumed the vehicle has received the attention outlined in section IV.

30. ENGINE.

a. Diagnosing Troubles. Determine troubles in a general way first as follows:

(1) CHECK MECHANICAL CONDITION. Check for mechanical trouble such as broken or deficient parts in engine or cylinder compression.

(2) CHECK IGNITION SYSTEM. Remove spark plug wire at a plug. Hold terminal end of wire about 1/4 inch from a metal part of engine, and check for a good spark by having someone turn ignition switch on and operate cranking motor. If no spark is obtained, check ammeter operation to determine condition of ignition primary circuit. Ammeter must show slight deflection from zero to discharge side (with lights off) when cranking motor is operated and ignition switch is on. If ammeter drops to zero when starting switch is pressed, starting system is defective, or battery is discharged.

(3) CHECK FUEL SYSTEM. Operate priming lever on rear side of fuel pump; to determine if fuel is reaching carburetor. Resistance to operation indicates carburetor is empty or no fuel; no resistance indicates carburetor is full. A flooded carburetor and engine may prevail so the spark plugs are shorted.

b. **Cranking Motor Will Not Crank Engine.**

(1) AMMETER DROPS TOWARD ZERO WHEN STARTING SWITCH IS PRESSED.

Possible Cause	Possible Remedy
Battery discharged.	Replace or charge battery (par. 97).
Battery terminals or ground cables loose or corroded.	Remove and clean.
Cranking motor drive gear jammed in flywheel teeth.	Rock vehicle backwards or loosen cranking motor (par. 89).
Excessive engine friction due to seizure or improper oil.	Change oil to proper grade (par. 18); if seizure has occurred, report to higher authority.

(2) AMMETER REMAINS UNCHANGED WHEN STARTING SWITCH IS PRESSED.

Battery cable terminal corroded or broken.	Clean or replace.
Poor starting switch contacts.	Replace switch (par. 90).

(3) CRANKING MOTOR RUNS BUT FAILS TO CRANK ENGINE WHEN SWITCH IS PRESSED.

Cranking motor gear does not engage flywheel.	Remove cranking motor and clean gear (par. 89).
Cranking motor or drive gear faulty.	Replace cranking motor (par 89).

c. **Engine Will Not Start.**

(1) NO SPARK.

(a) Ammeter Shows No Discharge (Zero Reading) with Ignition Switch "ON."

Ignition switch partly on.	Turn on fully.
Ignition switch faulty.	Replace switch (par. 68).

Passible Cause	Possible Remedy
Ignition primary wires, or cranking motor cables broken, or connections loose.	Repair or replace and tighten
Ignition coil primary winding open.	Replace coil (par. 66).
Distributor points burned, pitted, or dirty.	Clean or replace and adjust (par, 64).
Distributor points not closing.	Clean and adjust; put one drop of oil on arm post (par. 63).
Loose or corroded ground or battery cable connections.	Clean or replace and tighten.
Open circuit in suppression filter.	Test for trouble by removing ignition switch and coil wires, and connect together; if filter is faulty, report to higher authority.

(b) Ammeter Reading Normal.

High tension wire from coil to distributor broken, grounded, or out of terminals.	Repair or replace (par. 69).
Short-circuited secondary circuit in coil.	Replace coil (par. 66).
Short-circuited condenser.	Replace condenser (par. 64).
Short-circuited or burned distributor cap or rotor.	Replace part (par. 64).
Spark plugs, distributor cap, or wires wet (shorted).	Dry and clfaa thoroughly.
Spark plug gaps wrong.	Reset gaps (par. 67).
Ignition timing incorrect.	Set timing (par. 65).
Ignition wires installed wrong in distributor cap.	Put in proper places (par. 69).

(c) Ammeter Indicates Abnormal Discharge.

Short-circuited wire between ammeter and ignition switch or coil.	Repair or replace wire,
Short-circuited primary winding in ignition coil.	Install new coil (par. 66).
Radio filter short-circuited.	Disconnect temporarily, and report to higher authority.
Short-circuited condenser or broken lead.	Repair lead or replace condenser (par. 64).

TROUBLE SHOOTING

Possible Causa	Possible Remedy
Distributor points not opening.	Clean or replace and adjust (par. 63).
Distributor does not operate cam to open points.	Report to higher authority.
(2) WEAK SPARK.	
Distributor points pitted or burned.	Clean or replace and adjust (par. 64).
Distributor condenser weak.	Replace (par. 64).
Ignition coil weak.	Replace (par. 66).
Primary wire connections loose.	Tighten.
High tension or spark plug wires or distributor cap wet.	Dry thoroughly.
High tension or spark plug wires or distributor cap damaged.	Replace (par. 69).
Distributor rotor burned or broken.	Replace (par. 64).
(3) GOOD SPARK.	
Fuel tank empty.	Refill tank (par. 75).
Dirt or water in carburetor or float stuck.	Report to higher authority.
Carburetor and engine flooded by excessive use of choke.	Pull out throttle; crank engine with motor; when engine starts, regulate throttle; leave choke control "IN."
Choke control not operating properly.	Adjust (par. 72).
Fuel does not reach carburetor.	Check for damaged or leaky lines; air leak into line between tank and fuel pump.
Dirt in fuel lines or tank.	Disconnect drain tank and blow out lines.
Fuel line pinched.	Repair or replace.
Fuel strainer clogged.	Dismantle and clean (par. 76).
Fuel pump does not pump.	Clean screen; replace pump if inoperative (par. 74).
Lack of compression.	Report to higher authority.
(4) BACKFIRING.	
Ignition out of time.	Retime (par. 65).
Spark plug wires in wrong places in distributor cap or at spark plugs.	Install in proper places (par. 69).

Possible Cause	Possible Remedy
Distributor cap cracked or shorted.	Replace (par. 64).
Valve holding open—due to lack of compression.	Report to higher authority.

d. Engine Runs but Backfires and Spits

Possible Cause	Possible Remedy
Overheated engine.	Check (subpar. 1 below).
Improper ignition timing.	Reset (par. 65).
Spark plug wires in wrong place in distributor cap.	Install in proper places (par. 69).
Dirt or water in carburetor.	Clean and adjust (par. 72).
Carburetor improperly adjusted.	Check idle adjustment (par. 72).
Carburetor float level low.	Report to higher authority.
Valve sticking or not seating properly, burned, or pitted.	Report to higher authority.
Excessive carbon in cylinders.	Remove carbon (par. 54).
Valve springs weak.	Report to higher authority.
Heat control valve not operating.	Free-up and check thermostat spring position (par. 53).
Fuel pump pressure low.	Clean screen; replace pump, if faulty (par. 74).
Fuel strainer clogged.	Dismantle and clean (par. 76).
Partly clogged or pinched fuel line.	Clean or repair.
Intake manifold leak.	Check gaskets (par. 52).
Distributor capcracked or shorted.	Replace (par. 64).

e. Engine Stalls on Idle.

Possible Cause	Possible Remedy
Carburetor throttle valve closes too far, or idle mixture incorrect.	Adjust (par. 72).
Carburetor choke valve sticks closed.	Free-up and lubricate.
Dirt or water in idle passages of carburetor.	Replace carburetor (par. 72).
Air leak at intake manifold.	Tighten manifold stud nuts or replace gaskets (par. 52).
Heat control valve faulty.	Free-up and adjust (par 53).
Spark plugs faulty, gaps incorrect.	Clean or replace, set gaps (par. 67).
Ignition timing too early.	Reset (par. 65).
Low compression.	Report to higher authority.

TROUBLE SHOOTING

Possible Cause	Possible Remedy
Water leak in cylinder head or gasket.	Replace gasket, or report cylinder head leak to higher authority.
Crankcase ventilator valve stuck open.	Clean (par. 59).

f. Engine Misfires on One or More Cylinders.

Dirty spark plugs.	Clean and adjust or replace (par. 67).
Wrong type spark plugs.	Replace with correct type (par. 67).
Spark plug gap incorrect.	Reset gap (par. 67).
Cracked spark plug porcelain.	Replace spark plug (par. 67).
Spark plug or distributor suppressors faulty.	Replace (par. 67).
Spark plug wires grounded.	Replace.
Spark plug wires in wrong places in cap or at spark plugs.	Install correctly (par. 69).
Distributor cap or rotor burned or broken.	Replace (par. 64).
Valve tappet holding valve open.	Service (par. 56).
Compression poor—valve trouble.	Report to higher authority.
Leaky cylinder head gasket.	Replace gasket (par. 54).
Cracked cylinder block or broken valve tappet or tappet screw.	Report to higher authority.

g. Engine Does Not Idle Properly—(Erratic)

Ignition timed too early.	Reset (par. 65).
Dirty spark plugs or gaps too close.	Clean and adjust (par. 67).
Ignition coil or condenser weak.	Replace (par. 66).
Distributor points sticking, dirty or improperly adjusted.	Adjust or replace (par. 64).
Distributor rotor or cap cracked or burned.	Replace (par. 64).
Weak or broken valve spring.	Report to higher authority.
Leaky cylinder head gasket.	Replace (par. 54).
Uneven cylinder compression.	Report to higher authority.
High tension or spark plug wires leaky—cracked insulation.	Replace.
Dirt or water in carburetor, or float level incorrect.	Report to higher authority.
Carburetor adjustment or choke not set right.	Adjust (par. 72).

Possible Cause	Possible Remedy
Fuel pump pressure low.	Clean screen; replace pump (par. 74).
Crankcase ventilator valve leaks.	Clean (par. 59).
Leaky intake manifold.	Tighten manifold stud nuts or replace gaskets (par. 52).

Engine Misses On Acceleration.

Dirty spark plugs or gaps too wide.	Clean and adjust (par. 67).
Wrong type spark plug.	Replace (par. 67).
Ignition coil or condenser weak.	Replace (par. 66).
Distributor breaker points sticking, dirty or improperly adjusted.	Adjust or replace (par. 64).
Distributor cap or rotor cracked or burned.	Replace (par. 64).
Distributor cap, spark plugs or wire wet or dirty.	Clean and dry thoroughly.
High tension or spark plug wires leaky—cracked insulation.	Replace (par. 69).
Carburetor choke not adjusted.	Adjust (par. 72).
Carburetor accelerating pump system faulty, dirt in metering jets or float level incorrect.	Report to higher authority.
Fuel pump faulty—lack of fuel.	Clean screen; replace faulty pump (par. 74).
Air cleaner dirty.	Clean and reoil (par. 73).
Heat control valve faulty.	Check and adjust (par. 53).
Valves sticking—weak or broken valve springs.	Report to higher authority.
Overheated engine.	Check (subpar. I below).
Fuel strainer clogged.	Dismantle and clean (par. 76).

i. Engine Misses at High Speeds.

Distributor points sticking, adjusted too wide or burned.	Clean and adjust (par. 64).
Weak distributor arm spring.	Replace (par. 64).
Incorrect type of spark plugs.	Replace (par. 67).
Excessive play in distributor shaft bearing.	Replace distributor (par. 64).
Spark plugs faulty, dirty or incorrect gap.	Clean, adjust or replace (par. 67).
Weak ignition coil or condenser.	Replace (par. 66).

TROUBLE SHOOTING

Possible Cause	Possible Remedy
Valves sticking—weak or broken springs.	Report to higher authority.
Fuel supply lacking at carburetor.	Check fuel system (par. 71 a).
Heat control valve faulty.	Free-up and adjust (par. 53).
Air cleaner dirty.	Clean and reoil (par. 73).
Carburetor metering rod incorrectly set.	Report to higher authority.

j. **Engine Pings (Spark Knock).**

Ignition timing early.	Reset (par. 65).
Distributor automatic spark advance stuck in advance position or spring broken.	Replace distributor (par. 64).
Overheated engine.	Check (subpar. 1 below).
Excessive carbon deposit in cylinders.	Remove cylinder head and clean (par. 54).
Heat control valve faulty.	Free-up and adjust (par. 53).
Wrong type spark plug.	Replace (par. 67).
Old or incorrect fuel.	Drain and use correct fuel (par. 3).

k. **Engine Lacks Power.**

Ignition timing late.	Reset (par. 65).
Ignition system faulty.	Check (subpar. c above).
Old or incorrect fuel.	Use correct gasoline.
Leaky gaskets.	Replace.
Engine overheated.	Check (subpar. 1 below).
Excessive carbon formation.	*Remove cylinder head and dean* (par. 54).
Engine too cold.	Test thermostat (par. 85); in cold weather, cover radiator.
Insufficient oil or improper grade.	Use correct grade (par. 18).
Oil system failure.	Report to higher authority.
Air cleaner dirty.	Clean; change oil in reservoir (par. 73).
Spark plug gaps too wide.	Reset (par. 67).
Choke valve partially closed or throttle does not open fully.	Adjust (par. 72).
Manifold heat control inoperative.	Check valve operation; see that spring is in proper position (par. 53).

802011 O—48——6

Possible Cause	Possible Remedy
Exhaust pipe, muffler or tail pipe damaged or clogged.	Service or replace (par. 78).
Low compression—broken valve springs or sticking valves or improper tappet adjustment.	Report to higher authority.
Lack of fuel.	Clean filter (par. 76) check fuel pump (par. 74) check carburetor for water or dirt (par. 72).

I. Engine Overheats.

Possible Cause	Possible Remedy
Cooling system deficient.	Water low; air flow through radiator core restricted, clean from engine side; clogged core, clean or replace radiator (par. 81).
Radiator or water pump leaky.	Replace (par. 82).
Leaky cylinder head gasket.	Tighten or replace gasket (par. 54).
Damaged or deteriorated hose or fan belt.	Replace (par. 83).
Loose fan belt.	Adjust, or generator brace not hooked (par. 83).
Cylinder block, head or core hole plugs leaky.	Report to higher authority.
Ignition timing incorrect.	Reset (par. 65).
Damaged muffler; bent or clogged exhaust pipe.	Service or replace (par. 78).
Excessive carbon in cylinders.	Remove cylinder head and clean (par. 54).
Insufficient oil or improper grade.	Use correct grade (par. 18).
Air cleaner restricted.	Clean and renew oil (par. 73).
Inoperative thermostat or radiator cap.	Replace (par. 85).
Ignition system faulty.	Check (subpar. c above).
Water pump impeller broken.	Replace pump (par. 82).
Poor compression or valve timing wrong.	Report to higher authority.
Oil system failure (clogged screen).	Check (subpar. p below).

m. Low Fuel Mileage.

Possible Cause	Possible Remedy
High engine speeds (unnecessary and excessive driving in lower gear range).	Correct driving practice.

TROUBLE SHOOTING

Possible Cause	Possible Remedy
Air cleaner clogged.	Clean and renew oil (par. 73).
Carburetor float level too high. Metering rod, accelerating pump not properly adjusted.	Report to higher authority.
Fuel line leaks.	Tighten or replace.
Overheated engine.	Check (subpar. 1 above).
Carburetor parts worn or broken.	Replace carburetor (par. 72).
Fuel pump pressure too high or leaky diaphragm.	Replace fuel pump (par. 74).
Engine running cold.	Check thermostat (par. 85) ; cover radiator.
Heat control valve inoperative.	Free-up and put spring on bracket (par. 53).
Choke partially closed.	Adjust (par. 72).
Ignition timed wrong.	Reset (par. 65).
Spark advance stuck.	Replace distributor (par. 64).
Leaky fuel pump bowl gasket.	Replace gasket (par. 74).
Low compression.	Report to higher authority.
Carburetor controls sticking.	Free-up and lubricate.
Engine idles too fast.	Adjust carburetor throttle stop screw (par. 72).
Spark plugs dirty.	Clean or replace (par. 67).
Weak coil or condenser.	Replace (par. 64).
Clogged muffler or bent exhaust pipe.	Service or replace (par. 78).
Loose engine mountings permitting engine to shake and raise fuel level in carburetor.	Tighten; if damaged replace.

n. . Low Oil Mileage.

Possible Cause	Possible Remedy
High engine speeds or unnecessary and excessive driving in low gear ranges.	Correct driving practice.
Oil leaks.	Replace leaky gaskets.
Improper grade or diluted oil.	Use new oil of proper grade (par. 18).
Overheating of engine causing excessive temperature and thinning of oil.	Check (subpar. 1 above).
Oil filter clogged.	Clean; replace element (par. 58).

Possible Cause	Possible Remedy
Faulty pistons, or rings or rear bearing oil return clogged; excessive clearance of intake valves in guides; cylinder bores worn (scored, out-of-round, tapered); excessive bearing clearance; misalined connecting rods.	Report to higher authority.
o. Poor Compression.	
Incorrect tappet adjustment.	Adjust (par. 56).
Leaky, sticking or burned valves; sticking tappets; valve springs weak or broken; valve stems and guides worn; piston ring grooves worn or rings worn, broken or stuck; cylinders scored or worn excessively.	Report to higher authority.
p. Low Oil Pressure.	
Insufficient oil supply.	Check oil level.
Improper grade of oil or diluted oil foaming at high speeds.	Change oil; check crankcase ventilator (par. 59); check for water in oil by inspecting dip stick.
High oil temperature causing oil to be thin.	Check (subpar. 1 above).
Oil too heavy (funneling in cold weather).	Dilute engine oil (par. 18).
Floating oil intake loose or gasket leaky.	Renew gasket, tighten (par. 57).
Oil screen clogged.	Remove oil pan and clean screen (par. 57).
Oil leak causing lack of oil.	Inspect and service.
Faulty oil pump or pressure regulator valve stuck or spring broken.	Report to higher authority.-
Oil filter restriction hole too large.	Replace oil filter (par. 58).
Oil pressure too high.	Faulty oil pump regulator valve stuck closed or improperly adjusted, report to higher authority.
q. Faulty Valves.	
Incorrect tappet adjustment.	Adjust tappets (par. 56).
Other valve troubles.	Report to higher authority.

r. Abnormal Engine Noises.

Possible Causa	Possible Remedy
Loose fan, fan pulley or belt, heat control valve, or noisy generator brush.	Tighten or service.
Leaky intake or exhaust manifold or gaskets, cylinder head gasket or spark plug.	Replace or tighten (pars. 52 and 54).
Overheated engine; clogged exhaust system.	Remove obstruction from muffler tail pipe. Check (subpar. 1 above).
Other abnormal engine noises.	Report to higher authority.

31. CLUTCH.

a. Clutch Slips.

Improper pedal adjustment.	Adjust pedal free travel (par. 109).
Release linkage binding.	Free-up and lubricate.
Clutch facings burned or worn, torn loose from plate, or oil-soaked.	Replace clutch driven plate (pars. 110 and 111).
Weak pressure spring.	Report to higher authority.
Sticking pressure plate.	Report to higher authority.

b. Clutch Grabs or Chatters.

Control linkage binding.	Free-up and lubricate.
Loose engine mountings.	Tighten.
Engine stay cable not adjusted.	Adjust; just taut.
Facings burned, worn, or loose on driven plate; driven plate crimped or cushion flattened out, worn, or binding on splined shaft.	Replace clutch driven plate (pars. 110 and 111).
Pressure plate or flywheel face scored or rough; pressure plate broken; improper clutch lever (finger) adjustment; excessive looseness in power train.	Report to higher authority.

c. Clutch Drags.

Too much pedal play.	Adjust pedal free play (par. 109).
Driven plate warped; facings torn or loose.	Replace clutch driven plate (pars. 110 and 111).
Pressure plate warped or binds in bracket; improper finger adjustment; excessive friction in flywheel bushing.	Report to higher authority,

d. Clutch Rattles.

Possible Cause	Possible Remedy
Clutch pedal return spring is broken or disconnected.	Replace or connect.
Release fork loose on ball stud.	Adjust clutch pedal free travel to 3/4 inch (par. 109).
Driven plate springs broken. Worn release bearing.	Replace (pars. 110 and 111).
Worn pressure plate or broken return springs at driving lugs; worn driven plate hub on splined shaft; worn release bearing; fingers improperly adjusted; pilot bushing worn in flywheel.	Report to higher authority.

32. FUEL SYSTEM.

a. Fuel Does Not Reach Carburetor.

No fuel in tank.	Fill tank.
Fuel filter clogged.	Service fuel filter (par. 76).
Fuel pump inoperative.	Replace.
Fuel line air leak between tank and fuel pump.	Locate and correct.
Fuel line clogged.	Disconnect and blow out lines.
Fuel tank cap not functioning.	Replace cap.

b. Fuel Reaches Carburetor but Does Not Enter Cylinders.

Choke does not close.	Free-up and lubricate; inspect for proper operation.
Fuel passages in carburetor clogged.	Replace carburetor (par. 72).
Carburetor float valve stuck closed.	Report to higher authority.

c. Low Fuel Mileage.

Engine at fault.	Check (par. 30 m above),
Lubricant in power train too heavy.	Use correct lubricant (par. 18).
Tires improperly inflated.	Inflate (par. 3).
Vehicle overloaded.	Reduce to 500 pounds if possible.

d. Low Fuel Pressure.

Air leak in fuel lines.	Tighten connections; repair if damaged; hand-tighten fuel pump dome nut.

TROUBLE SHOOTING

Possible Cause	Possible Remedy
Fuel pump faulty; diaphragm broken; valves leaky; linkage worn.	Replace fuel pump (par. 74).
Fuel lines clogged.	Clean or replace lines.

e. Engine Idles Too Fast.

Improper carburetor throttle adjustment.	Adjust throttle stop screw (par. 72).
Carburetor control sticking.	Free-up and lubricate.
Control return spring weak.	Replace.

f. Fuel Gage Does Not Register.

Loose wire connection at instrument panel or tank units.	Tighten connection.
Instrument panel unit or tank unit inoperative.	Replace (pars. 75 and 77).

33. INTAKE AND EXHAUST SYSTEMS,

a. Intake System.

Leaky gaskets, sand hole or crack in manifold.	Replace (par. 52).
Leaky crankcase ventilator valve.	Replace (par. 59).

b. Exhaust System.

Leaky gaskets, sand hole or crack in manifold.	Replace (par. 52).
Exhaust pipe and connections loose or leaking.	Service and/or replace (par. 78).
Muffler leaks or rattles.	Replace (par. 78).
Exhaust system or muffler restricted; exhaust pipe kinked or tail pipe plugged.	Service or replace parts.
Heat control valve inoperative, causing miss on acceleration or slow warm-up.	Free-up; install spring in place on bracket (par. 53).

34. COOLING SYSTEM.

a. Overheating.

Abnormal conditions.	Check (par. 30 1).

b. Loss of Cooling Solution.

Loose hose connection.	Tighten.
Damaged or deteriorated hose.	Replace.

Possible Cause	Possible Remedy
Leaky radiator.	Replace (par. 81).
Radiator cap inoperative.	Replace.

c. Engine Running Too Cool.

Possible Cause	Possible Remedy
Thermostat stuck open.	Replace (par. 85).
Low air temperatures.	Cover radiator; refer to operation under unusual conditions (par. 7).

d. Noises.

Possible Cause	Possible Remedy
Frayed or loose fan belt.	Replace or adjust (par. 83).
Water pump faulty.	Replace (par. 82).
Fan blades striking.	Aline blades.

35. IGNITION SYSTEM.

a. Ignition System Troubles.

Possible Cause	Possible Remedy
No spark.	Refer to paragraph 30 c (1).
Weak spark.	Refer to paragraph 30 c (2).
Timing incorrect.	Retime ignition (par. 65); refer to paragraph 30 j for other causes.
Moisture on distributor wires, coil, or spark plugs.	Dry and clean thoroughly with cloth dampened with carbon tetrachloride.
Ignition switch "OFF."	Turn "ON" fully.
Ignition switch does not make contact.	Replace switch (par. 68).
Primary or secondary wiring loose, broken, or grounded.	Service.
Primary or secondary wiring wrong.	Check against wiring diagram (par. 62 and fig. 30); install secondary wires correctly in distributor cap and on spark plugs.
Ground strap connections (engine to frame) loose or dirty.	Clean and tighten.
Coil faulty.	Refer to subparagraph b below.
Distributor faulty.	Refer to subparagraph c below.
Spark plug or distributor suppressors faulty.	Replace (par. 67).
Filter unit open or grounded.	Replace filter (par. 69).

b. Ignition Coil Troubles.

Possible Cause	Possible Remedy
Connections loose; dirty or broken external wire; wet.	Clean and tighten or repair; dry thoroughly.
Coil internal fault.	Replace coil (par. 66).

c. Distributor Troubles.

Distributor breaker points dirty or pitted; gap incorrect.	Clean or replace and adjust (par. 64).
Distributor breaker point arm spring weak.	Replace breaker point arm (par. 64).
Distributor breaker points stuck open.	Free-up and lubricate arm on post.
Distributor automatic advance faulty.	Lubricate and free up; if "frozen" replace distributor (par. 64).
Distributor cap or rotor shorted, cracked, or broken.	Replace.
Distributor rotor does not turn.	Report to higher authority.
Distributor cap cracked or shorted.	Replace cap (par. 64).
Condenser or lead wire faulty.	Replace condenser (par. 64).

d. Spark Plug Troubles.

Cracked, broken, leaky, or improper type.	Replace spark plug (par. 67).
Spark plug wires installed on wrong plugs, or in distributor cap.	Install in correct place (par. 69).
Spark plugs dirty; gaps incorrect.	Clean or replace; set gaps (par. 67).
Spark plug porcelain cracked or broken.	Replace plug.
Spark plugs wrong type.	Replace with correct type (par. 67).

36. STARTING AND GENERATING SYSTEMS.

a. Cranking Motor Troubles.

(1) CRANKING MOTOR CRANKS ENGINE SLOWLY.

Engine oil too heavy.	Change to proper seasonal grade (par. 18).
Battery low.	Replace or recharge (par. 97).
Battery cell shorted.	Replace battery (par. 97).

Possible Cause	Possible Remedy
Battery connections corroded, broken, or loose; or engine ground strap to frame connections dirty or loose.	Clean and tighten or replace (par. 97).
Dirty commutator.	Clean (par. 89).
Poor brush contact.	Free-up brush or replace cranking motor (par. 89).
Cranking motor internal fault.	Replace cranking motor.
Starting switch faulty.	Replace switch (par. 90).

(2) CRANKING MOTOR DOES NOT CRANK ENGINE.

Engine oil too heavy.	Change to proper seasonal grade (par. 18).
Cranking motor, starting switch or cables faulty; loose connections.	Replace; tighten loose connections.

b. Generator Troubles.

(1) No OUTPUT.

Generator faulty.	Replace generator (par. 93).
Filter unit or suppressors faulty.	Replace (par. 93).
Regulator faulty.	Replace (par. 94).

(2) Low OR FLUCTUATING OUTPUT.

Loose fan belt.	Adjust (par. 83); generator brace not hooked (par. 4 b (8)).
Poor brush contact, weak brush springs; worn commutator; broken or loose connections.	Replace generator (par. 93).
Dirty commutator.	Clean (par. 93).
Regulator faulty.	Replace (par. 94)
Loose or dirty connections in charging circuit.	Clean and tighten.
Ground strap (engine to frame) broken.	Replace.
Filter unit faulty.	Replace (par. 93).

(3) EXCESSIVE OUTPUT.

Short circuit between field coil and armature leads.	Replace generator (par. 93).
Regulator faulty.	Replace regulator (par. 94).

(4) NOISY.

Loose pulley or generator mounting.	Tighten.

TROUBLE SHOOTING

Possible Cause	Possible Remedy
Faulty bearings, improperly seated brushes, or armature rubbing on field poles.	Replace generator (par. 93).

c. Generator Regulator Troubles.

Loose connections or mounting.	Clean and tighten.
Regulator internal defect.	Replace regulator (par. 94).

37. TRANSMISSION.

a. Excessive Noise.

Incorrect driving practice.	Correct practice (par. 5).
Insufficient lubricant.	Add lubricant (par. 18).
Incorrect lubricant.	Use correct lubricant (par. 18).
Gears or bearings broken or worn; shift fork bent; gears worn on splines.	Replace transmission (pars. 115 and 116).
Overheated transmission.	Check lubricant grade and supply (par. 18).

b. Hard Shifting.

Clutch fails to release.	Adjust clutch pedal free travel (par. 109).
Clutch driven plate binds on splines, or pressure plate faulty.	Report to higher authority.
Gearshift binding in housing.	Lubricate and free-up.
Shift rods binding in case.	Report to higher authority.
Transmission loose on bell housing.	Tighten.
Clutch shaft pilot binding in bushing case or shift housing damaged.	Report to higher authority.

c. Slips Out of Gear.

Weak or broken poppet spring.	Report to higher authority.
Interlock plunger not in place.	Install plunger (par. 116).
Transmission gears or bearings worn.	Replace transmission (pars. 115 and 116).
Shift fork bent, causing partial gear engagement.	Report to higher authority.
Transmission loose on bell housing.	Tighten.
Damaged bell housing.	Report to higher authority.

d. Loss of Lubricant.

Worn or damaged seals or gaskets.	Report to higher authority.

Possible Cause	Possible Remedy
Overfilled with lubricant.	Drain to proper level.
Loose bolts and screws.	Tighten.

38. TRANSFER CASE.

a. Slips Out of Gear.

Possible Cause	Possible Remedy
Shift rod poppet spring weak or broken; gears not fully engaged; shift fork bent; end play in sliding gear shaft.	Report to higher authority.
Parts damaged or worn.	Replace transfer case (pars. 119 and 120).

b. Hard Shifting.

Possible Cause	Possible Remedy
Improper driving practice.	Use correct procedure (par. 5).
Lack of lubrication.	Replenish supply.
Shift lever seizing on shaft.	Lubricate and free-up.
Shift rod tight in case; poppet scored or stuck; shift fork bent, or parts worn or damaged.	Report to higher authority.
Low or uneven tire pressures; odd tires on (front and rear) wheels.	Service.

c. Oil Leaks.

Possible Cause	Possible Remedy
Leaks at gaskets or seals.	Report to higher authority.
Lubricant level too high.	Reduce to correct level.
Vent on top of unit clogged.	Clean.

d. Excessive Noise.

Possible Cause	Possible Remedy
Insufficient lubricant.	Replenish supply.
Incorrect lubricant.	Drain and refill with correct lubricant (par. 18).
Gears or bearings worn, improperly adjusted, or damaged.	Replace transfer case (pars. 119 and 120).

e. Overheats.

Possible Cause	Possible Remedy
Insufficient lubricant.	Replenish supply.
Vent on top of unit clogged.	Clean.
Bearings adjusted too tight.	Report to higher authority.

f. Backlash.

Possible Cause	Possible Remedy
Universal joint yoke loose on output shaft.	Report to higher authority.

TROUBLE SHOOTING

Possible Cause	Possible Remedy
Transfer case loose on transmission or snubbing rubber.	Tighten.
Parts worn or damaged.	Report to higher authority

39. PROPELLER SHAFTS.

a. Excessive Vibration or Noise.

Possible Cause	Possible Remedy
Foreign material around shaft.	Clean out.
Universal joints not in same plane.	Match arrows on joint and propeller shaft (par. 125).
Lack of lubricant.	Lubricate (par. 18).
Universal joint parts worn, or propeller shaft sprung.	Replace shaft.

b. Universal Joint Leaks.

Possible Cause	Possible Remedy
Overfilled.	Lubricate correctly (par. 18).
Oil seals leak.	Report to higher authority.
Lubricant fitting leaks.	Replace fitting.

40. FRONT AXLE.

Possible Cause	Possible Remedy
a. Steering trouble.	Refer to paragraph 45.
b. Noisy gears or backlash.	Report to higher authority.
c. Damaged axle.	Replace axle (pars. 136 and 137).
d. Abnormal tire wear.	Inflate tires (par. 13 b (13)) (do not use front wheel drive except where needed); correct toe-in; report to higher authority incorrect caster or camber.
e. Lubrication leaks.	Replace steering knuckle oil seals; for other remedies refer to paragraph 41 c.

41. REAR AXLE.

Possible Cause	Possible Remedy
a. Noisy gears or backlash.	Report to higher authority.
b. Damaged axle.	Replace axle (par. 145).
c. Lubrication leaks.	Drain excessive lubricant; clean housing vent; replace wheel bearing grease seals; remove excessive grease in wheel hubs; tighten or replace housing cover gasket.

42. BRAKE SYSTEM.

a. All Brakes Drag.

Possible Cause	Possible Remedy
Improper pedal adjustment.	Adjust brake pedal free travel (par. 148).
Clogged master cylinder port.	Replace (par. 150).
Brake pedal return spring broken or weak.	Replace,
Brakes improperly adjusted.	Adjust (par. 148).
Rubber parts swollen from use of mineral oil in brake fluid.	Report to higher authority,

b. One Brake Drags.

Brake shoe adjustment faulty.	Adjust (par. 148).
Brake shoe anchor pin tight in shoes.	Free-up and lubricate lightly,
Brake shoe return spring broken or weak.	Replace,
Brake hose clogged or pinched.	Replace.
Loose or damaged wheel bearings.	Adjust or replace (pars. 128 and 141).
Wheel cylinder pistons or cups faulty.	Replace wheel cylinder (par. 150).

c. One Brake Grabs (Vehicle Pulls to One Side).

Tires underinflated.	Inflate tires (par. 13).
Tires worn unequally.	Replace.
Insufficient brake shoe clearance or brake anchor pin adjustment faulty.	Adjust (par. 148).
Axle spring clips or brake backing plate loose.	Tighten,
Brake shoes binding on anchor pin.	Free-up and lubricate lightly,
Weak or broken shoe return spring.	Replace spring,
Grease or brake fluid on linings.	Correct leakage; clean up and install new shoes and lining assemblies.
Dirt imbedded in linings or rivet holes.	Clean with wire brush,
Drums scored or rough.	Replace drums and brake shoe and lining assemblies.

TROUBLE SHOOTING

Possible Cause	Possible Remedy
Primary and secondary brake shoes reversed in one wheel.	Change shoes to proper place and adjust brakes (par. 148).
Odd kinds of brake lining on opposite wheels.	Replace shoe and lining assemblies in both wheels.
Loose or broken wheel bearings.	Adjust or replace (pars. 128 and 141).
Obstruction in brake line.	Clean or replace tube (par. 152).

d. Severe Brake Action on Light Pedal Pressure,

Brake shoes improperly adjusted.	Adjust (par. 148).
Grease or brake fluid on linings.	Correct leakage, clean up and replace shoe and lining assemblies.
Loose brake shoe anchor.	Adjust and tighten (par. 148).
Improper linings.	Replace shoe and lining assemblies (par. 148).

e. Brakes Locked.

Brake pedal lacks free travel.	Adjust pedal free travel (par. 148).
Bleed hole in master cylinder clogged.	Replace master cylinder (par. 150).
Dirt in brake fluid.	Flush system (par. 151).
Wheel cylinder stuck.	Replace cylinder (par. 150).
Brakes frozen to drums (cold weather).	Break loose by driving vehicle.

f. Brakes Noisy or Chatter.

Brake lining worn out.	Replace shoe and lining assemblies (par. 148).
Grease or brake fluid on linings.	Correct leakage, clean up and replace shoe and lining assemblies (par. 148).
Improper adjustment of anchor bolts.	Adjust (par. 148).
Dirt imbedded in linings and rivet holes.	Clean with wire brush.
Improper or loose linings.	Replace shoe and lining assemblies (par. 148).
Brake shoes, drums, or backing plate distorted.	Straighten or replace.
Loose spring clips or shackles.	Tighten.

g. Excessive Pedal Travel.

Pottibls Cause	Possible Remedy
Normal lining wear.	Adjust brake eccentrics only (par. 148).
Lining worn out.	Replace shoe and lining assemblies (par. 148).
Brake not properly adjusted.	Adjust (par. 148).
Improper pedal adjustment.	Adjust (par. 148).
Brake line leaky or broken.	Locate and tighten or repair.
Low fluid level in master cylinder or air in brake system.	Fill master cylinder and bleed lines (par. 151).
Scored brake drums.	Replace (pars. 131 and 144).
Incorrect brake lining.	Replace with correct shoe and lining assemblies.
Pedal goes to floorboard (disconnected from master cylinder).	Connect or replace faulty part (par. 149).
Leaky piston cup in master or wheel cylinders.	Replace cylinder.

h. Excessive Pedal Pressure.

Grease or brake fluid on linings; worn or glazed lining.	Correct cause, clean up and replace shoe and lining assemblies (par. 148).
Warped shoes or improper brake linings.	Replace shoe and lining assemblies (par. 148).
Shoes improperly adjusted.	Adjust (par. 148).
Brake drums scored or distorted.	Replace damaged parts.
Improper brake fluid.	Clean system and fill with correct fluid.
Obstructed main brake line.	Locate and correct.

i. Spongy Brake Pedal Action.

Air or insufficient fluid in brake system.	Fill master cylinder and bleed lines (par. 149).
Brake anchor adjustment faulty.	Adjust (par. 148).

j. No Brakes-Pedal Will Pump Up.

Brake shoe clearance excessive.	Adjust brake eccentrics (par. 148).
Leaky master or wheel cylinder piston cup.	Replace cylinder,
Leaky brake line or hose.	Locate and tighten or replace.

k. Pedal Goes to Floor Slowly When Brakes Are Applied.

Poisible Cause	Possible Remedy
Leaky master cylinder piston cup.	Replace master cylinder (par. 149).
Leaky brake line or hose.	Tighten or replace part.

43. WHEELS, WHEEL BEARINGS, AND RELATED PARTS.

a. Wheel Troubles.

Wheel wobbles; bent.	Check mounting on hub; replace bent wheel.
Wheel loose on hub.	Tighten.
Wheel out of balance.	Remount tire correctly.
Wheel bearings run hot (pull vehicle to one side).	Adjust (pars. 128 and 141).
Wheels misalined.	Refer to paragraph 135.
Excessive or uneven tire wear.	Refer to paragraph 45.
Diameter of front tires not the same in size or wear.	Replace or match up.

44. SPRINGS AND SHOCK ABSORBERS

a. Broken Springs.

Improper handling of vehicle on rough terrain.	Use correct practice when possible (par. 5).
Overloaded vehicle.	Reduce load (par. 3).
Overlubricated springs.	Do not lubricate unless rusty.
Rebound clips off or out of place.	Service.
Shackles or pivot bolts too tight.	Free-up and lubricate.
Main leaf broken at end.	Replace spring (par. 156).
Axle clips loose (spring broken at center).	Keep clips tight.
Shock absorbers not adjusted correctly, lack fluid, or damaged.	Adjust or replace shock absorbers (par. 157).
Clutch or brakes grab,	Service.

b. Noisy Springs.

Worn shackles, pivot pins, or bushings.	Replace worn parts (par. 155).
Spring clips loose on axle or leaves.	Tighten.
Spring hangers loose on frame.	Report to higher authority.
Spring shackle bushing loose; inner spring eye opened up.	Replace spring (par. 156).

Possible Cause	Possible Remedy
No fluid in shock absorbers, or bushings worn out.	Replace (par. 157).

c. Bottomed Springs.

Overloaded vehicle.	Reduce load (par. 3).
Overlubricated springs.	Do not lubricate unless rusty.
Broken spring leaves.	Replace spring (par. 156).
Shock absorbers broken, lack fluid or proper adjustment.	Replace shock absorbers (par. 159).

d. Overflexible Springs.

Overlubrication causes springs to bottom.	Do not lubricate springs.
Shock absorbers not adjusted right, lack fluid, or are broken.	Service, adjust, or replace shock absorbers (par. 157).
Rebound clips damaged or lost.	Replace.
Broken spring.	Replace spring (par. 156).

e. Stiff Springs.

Rusted spring leaves.	Lubricate.
Shackle or pivot bolts too tight.	Free-up and lubricate.
Shock absorber adjustment not right.	Adjust (par. 157).

f. Noisy Shock Absorbers.

Rubber bushings worn out.	Replace bushing (par. 157).
Mounting bracket loose.	Report to higher authority.
Shock absorber faulty.	Replace (par. 157).

g. Shock Absorber Control Too Stiff or Too Soft.

Shock absorber adjustment wrong.	Adjust (par. 157).
Shock absorber damaged or lacks fluid.	Replace shock absorber (par. 157).

45. STEERING SYSTEM,

a. Steering Difficult.

Lack of lubrication.	Lubricate (par. 18).
Tire pressures low.	Inflate (par. 13).
Tight steering system connections.	Lubricate and adjust (par. 159).
Tight steering gear; misalined front wheels (caster or camber); or bent frame.	Report to higher authority.

Improper front wheel toe-in. Bent steering connecting parts.

Misalined steering gear mounting.

b. Wander or Weaving.

Improper toe-in

Improper camber or caster (axle twisted).

Front springs settled or broken.

Axle shifted (spring center bolt broken).

Loose or lost spring clips.

Loose or worn spring shackles or bolts.

Tire pressures uneven.

Steering system connections or king pin bearings not properly adjusted.

Loose wheel bearings. Faulty shock absorbers. Steering gear worn or out of adjustment.

Steering gear mounting loose.

Steering Pitman arm loose.

Possible Remedy

Adjust (par. 135).

Loose steering connections.

Spring clips or shackles loose.

Front axle loose on spring (broken spring center bolt).

Insufficient toe-in.

Improper caster or twisted axle. Steering gear worn, or adjustments too loose.

Loose wheel or king pin bearings.

d. High Speed Shimmy or

listed in subparagraph c above).

Tire pressures low or uneven.

Straighten or replace.

Adjust mounting.

Adjust (par. 135).

Report to higher authority.

Replace spring (par. 156). Replace part.

Tighten or replace.

Replace or tighten.

Inflate (par. 13).

Lubricate and adjust (par. 159).

Adjust (pars. 128 and 141).

Replace (par. 157).

Report to higher authority.

Tighten.

Tighten.

e. Low Speed Shimmy or Wobble.

Wheels and tires out of balance.

Adjust.

Adjust or replace.

Replace bolt.

Adjust (par. 135).

Report to higher authority.

Report to higher authority.

Adjust (par. 128).

Tight Wheel (Refer to remedies

Inflate (par. 13).

Check tire mounting; report other trouble to higher authority.

Possible Cause	Possible Remedy
Wheel run-out; tire radial run-out or wheel camber incorrect.	Report to higher authority.
Front springs settled or broken.	Replace spring (par. 156).
Bent steering knuckle arm.	Report to higher authority.
Shock absorbers not effective.	Adjust or replace.
Steering gear loose in frame.	Tighten.
Front springs too flexible.	Do not lubricate.
Worn spring bolts, shackles, or bushings.	Replace (par. 155).
Axle housing or frame damaged,	Report to higher authority.

e. Wheel Tramp (High Speed).

Wheels and tires out of balance.	Check tire mounting; report other trouble to higher authority.
Uneven tire wear.	Shift tires.
Shock absorbers ineffective.	Replace or adjust (par. 157).

f. Vehicle Pulls to One Side.

Tires not inflated evenly.	Inflate (par. 13).
Unequal caster or camber (bent axle).	Report to higher authority.
Odd size, or new and old tires on opposite front wheels.	Switch tires.
Tight wheel bearing.	Adjust (pars. 128 and 141).
Bent steering arm or connection.	Straighten or replace.
Brake drag.	Adjust brakes (par. 148).

g. Road Shock.

Tightness in steering connecting parts.	Adjust (par. 159).
Excessive spring flexibility.	Do not lubricate.
Loose wheel bearings.	Adjust.
Loose Pitman arm or mounting.	Tighten.
Looseness in steering gear.	Report to higher authority.
Shock absorbers out of adjustment or faulty.	Adjust or replace (par. 157).

h. Steering Dive.

Steering gear loose on frame.	Tighten.
Broken front spring leaves.	Replace.
Worn spring shackles, bushings or bolts.	Replace.

Possible Causa	Possible Remedy
Spring hangers loose on frame.	Report to higher authority.
Spring clips loose, broken, or lost.	Tighten or replace.
Spring center bolt broken and/or clips loose.	Replace.
Axle housing on frame damaged.	Report to higher authority.

1. Unequal Steering (Right and Left).

Pitman arm not installed in proper position on steering gear.	Remove and install in correct position (par. 162).
Drag link bent.	Straighten or replace.

46. BODY AND FRAME.

a. Body.

Worn or damaged seat cushion.	Replace.
Badly damaged fender, radiator guard, hood, fuel can rack, seats, top, or windshield.	Replace; report minor damage to higher authority.
Windshield wiper faulty.	Service or replace.

b. Frame.

Badly damaged bumpers and pintle hook.	Replace; report minor damage to higher authority.
Damaged frame.	Report to higher authority.

47. BATTERY AND LIGHTING SYSTEM

a. Battery.

(1) BATTERY DISCHARGED.

Battery solution level low.	Add distilled water to bring level above plates; check for cracked case.
Short in battery cell.	Replace battery (par. 97).
Generator not charging.	Check generator, fan belt and regulator (par. 92).
Loose or dirty connections; broken cables.	Clean and tighten connections; replace cables.
Excessive use of cranking motor.	Tune up engine; charge battery.
Idle battery, or excessive use of lights.	Replace or charge battery.

(2) BATTERY (OTHER TROUBLES.)

Overheated battery.	Check for short circuit or excessive generator charge.

Possible Cause	Possible Remedy
Case bulged or out of shape.	Check for overcharging and too tight hold-down screws.
b. Switch.	
Loose or dirty connections or broken wire.	Clean and tighten; replace broken wire.
Internal fault.	
c. Fuse (Circuit Breaker).	Replace switch.
Points dirty.	
Other troubles.	Clean.
d. Wiring.	
Loose or dirty connections, broken wire or terminal.	Replace fuse assembly (par. 104).
e. Lights do Not Light.	Clean, tighten or replace.
Switch not fully on.	Turn switch on fully.
Loose or dirty connection, or broken wire or terminal.	Clean and tighten; replace or repair wire or terminal.
Wiring circuit shorted or open.	Localize and repair.
Headlight, blackout driving light, tail or stop light burned out.	Replace lamp-unit (par. 96).
Blackout headlight burned out.	Replace lamp (par. 100).
f. Lights Dim.	
Loose or dirty connection or poor ground connection.	Clean and tighten.
Wire grounding.	Localize and replace (par. 98).
Poor switch contact.	Replace switch.
Headlight aim not right.	Adjust lights (par. 99).
g. Trailer Connection Trouble	
No current supply.	Tighten loose wires; connect to correct terminals.
h. Horn Troubles.	
Loose or dirty connections.	Clean and tighten.
Sounds continuously (short circuit in wiring between horn and horn button).	Replace wire.
Improper tone.	Adjust points; tighten cover or bracket screws; clean and tighten loose or dirty wiring connections.

Possible Cause	Possible Remedy
Internal defect.	Replace horn.
Battery low.	Charge or replace battery.

48. RADIO SUPPRESSION,

a. Radio Interference.

Possible Cause	Possible Remedy
Faulty ignition.	Check distributor, spark plugs, and suppressors. Tighten braided bonding straps. Tighten radiator and fender supporting bolts. Check high-tension insulation. Tighten loose wiring connections, or replace corroded distributor cap towers. Replace defective switches or gages (par. 178).
Faulty generator.	Tighten generator to regulator bond. Check for faulty commutator, brushes, or holders. If defective, replace generator. Check for discharged battery causing high charging rate (par. 178).
Erratic noises.	Tighten or clean loose or dirty lock washer ground. Install lock washers in correct position (par. 178).

49. INSTRUMENTS,

a. Faulty Instruments.

Possible Cause	Possible Remedy
Dirty or loose connections.	Clean and tighten.
Internal defects.	Replace instrument.
Broken speedometer cable.	Replace (par. 166).

Section XIII

ENGINE-DESCRIPTION, DATA, MAINTENANCE, AND ADJUSTMENT IN VEHICLE

Paragraph

50. DESCRIPTION AND TABULATED DATA.

a. Description. The engine (figs. 20 and 21) is of the conventional 4-cylinder, L-head, internal-combustion type. The engine with the clutch, transmission, and transfer case is built into a unit power plant which is mounted at four points in the chassis. For identification refer to paragraph 2 b.

b. Tabulated Data.

Type	L-head
Number of cylinders	4
Bore	31/8 in.
Stroke	43/8 in.
Piston displacement	134.2 cu in.
Compression ratio	6.48 to 1
Net horsepower	54 at 4,000 rpm
Compression	110 lb per sq in. at 185 rpm
SAE horsepower	15.63
Maximum torque	95 ft-lb
Firing order	1-3-4-2
Tappet clearance—intake and exhaust (hot or cold)	0.014 in.

51. ENGINE TUNE-UP.

a. Procedure.

(1) Perform preventive maintenance and corrective operations listed in paragraph 16.

(2) Remove spark plugs and clean. Adjust gaps (par. 67).

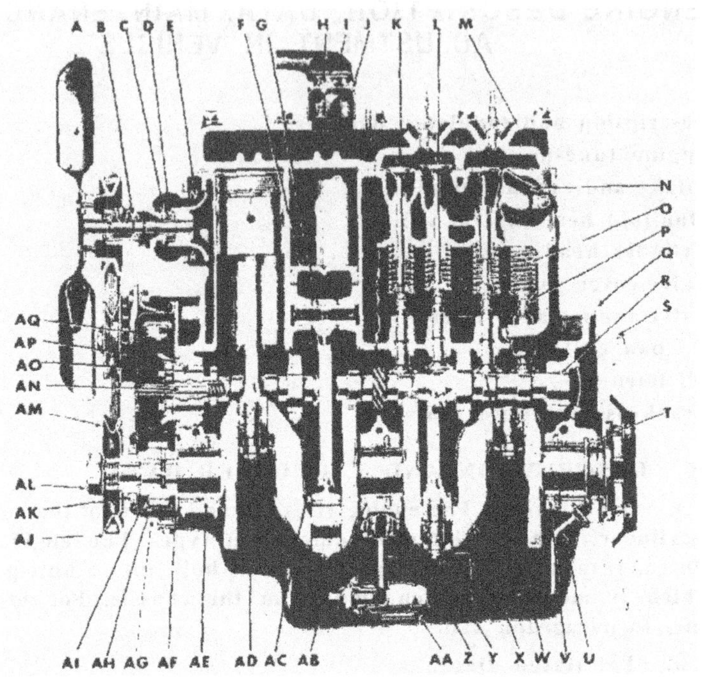

A FAN
B WATER PUMP BEARING AND SHAFT
C WATER PUMP SEAI WASHER
D WATER PUMP SEAL
E WATER PUMP IMPELLER
F PISTON
G PISTON PIN
H THERMOSTAT
I WATER OUTLET ELBOW
J THERMOSTAT RETAINER
K EXHAUST VALVE
L INLET VALVE
M CYLINDER HEAD
N EXHAUST MANIFOLD
O VALVE SPRING
P VALVE TAPPET ADJUSTING SCREW
Q ENGINE PLATE—REAR
R CAMSHAFT
S FLYWHEEL RING GEAR
T CRANKSHAFT PACKING—REAR
U CRANKSHAFT REAR BEARING DRAIN PIPE
V CRANKSHAFT BEARING—REAR LOWER
W VALVE TAPPET
X CRANKSHAFT
Y CONNECTING ROD CAP BOLT
Z OIL FLOAT SUPPORT
AA OIL FLOAT
AB CRANKSHAFT BEARING—CENTER LOWER
AC CONNECTING ROD
AD CONNECTING ROD BOLT NUT LOCK
AE CRANKSHAFT BEARING—FRONT LOWER
AF CRANKSHAFT THRUST WASHER
AG TIMING CHAIN COVER
AH TIMING CHAIN
AI CRANKSHAFT SPROCKET
AJ FAN BELT
AK CRANKSHAFT PACKING—FRONT END
AL STARTING CRANK NUT
AM FAN AND GENERATOR DRIVE PULLEY
AN CAMSHAFT THRUST PLUNGER
AO CAMSHAFT BUSHING—FRONT
AP CAMSHAFT THRUST WASHER
AQ CAMSHAFT SPROCKET

RA PD 305283

Figure 20—Sectional View of Engine

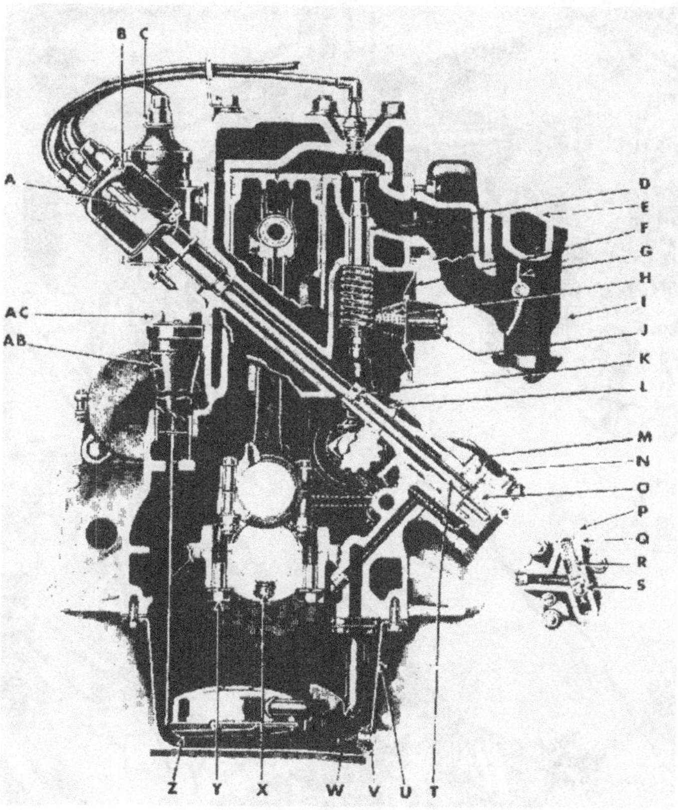

A DISTRIBUTOR OILER
B IGNITION DISTRIBUTOR
C IGNITION COIL
D EXHAUST VALVE GUIDE
E INTAKE MANIFOLD
F VALVE SPRING COVER
G HEAT CONTROL VALVE
H CRANKCASE VENTILATOR BAFFLE
I EXHAUST MANIFOLD
J CRANKCASE VENTILATOR
K DISTRIBUTOR SHAFT FRICTION SPRING
L OIL PUMP DRIVEN GEAR
M OIL PUMP ROTOR DISK
N OIL PUMP
O OIL PUMP PINION
P OIL RELIEF PLUNGER SPRING RETAINER
Q OIL RELIEF PLUNGER SPRING SHIMS
R OIL RELIEF PLUNGER SPRING
S OIL RELIEF PLUNGER
T OIL PUMP SHAFT AND ROTOR
U OIL PAN
V OIL PAN DRAIN PLUG
W OIL FLOAT SUPPORT
X CRANKSHAFT BEARING DOWEL
Y CRANKSHAFT BEARING CAP TO CRANKCASE SCREW
Z OIL FLOAT
AB OIL FILLER TUBE
AC OIL FILLER CAP AND LEVEL INDICATOR

RA PD 305284

Figure 21—Sectional View of Engine

Figure 22—Manifolds

(3) Test cylinder compression with gage. The gage must read more than 70 pounds, and the variation between cylinders remain less than 10 pounds. Normal compression is approximately 110 pounds per square inch at cranking speed. Report lack of compression to higher authority.

(4) Make sure that ground strap at engine left front support is in good condition and tight.

(5) Remove distributor cap and rotor. Check for cracks and leaks. Clean or replace breaker points and adjust (par. 64).

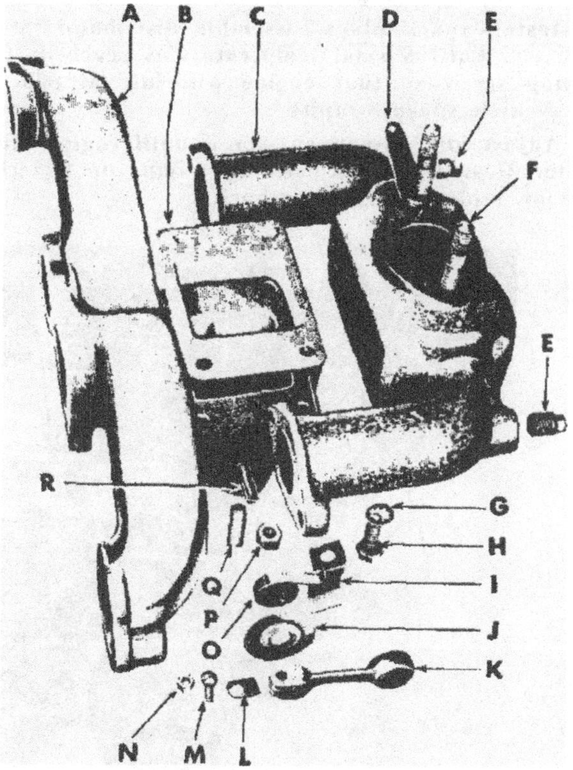

A EXHAUST MANIFOLD
B INTAKE TO EXHAUST MANIFOLD GASKET
C INTAKE MANIFOLD
D ACCELERATOR SPRING CLIP
E INTAKE MANIFOLD PLUG
F INTAKE MANIFOLD TO CARBURETOR STUD
G INTAKE TO EXHAUST MANIFOLD SCREW LOCKWASHER
H INTAKE TO EXHAUST MANIFOLD SCREW
I HEAT CONTROL VALVE BI-METAL SPRING STOP
J HEAT CONTROL VALVE BI-METAL SPRING WASHER
K HEAT CONTROL VALVE COUNTERWEIGHT LEVER
L HEAT CONTROL VALVE LEVER KEY
M HEAT CONTROL VALVE LEVER CLAMP SCREW
N HEAT CONTROL VALVE LEVER CLAMP SCREW NUT
O HEAT CONTROL VALVE BI-METAL SPRING
P EXHAUST PIPE TO EXHAUST MANIFOLD STUD NUT
Q EXHAUST PIPE TO EXHAUST MANIFOLD STUD
R HEAT CONTROL VALVE SHAFT

RA PD 334754

Figure *23—Manifolds, Disassembled*

(6) Check ignition timing (par. 65).

(7) Check valve tappet clearance (par. 56).

(8) Install spark plugs, assemble distributor, start engine, and allow to run until normal temperature is reached; then set throttle valve stop screw so that engine will idle at 600 revolutions per minute (vehicle speed 8 mph).

(9) Adjust idle adjustment screw until engine idles smoothly. If carburetor float level, accelerating pump, or metering rod require adjustment, report to higher authority.

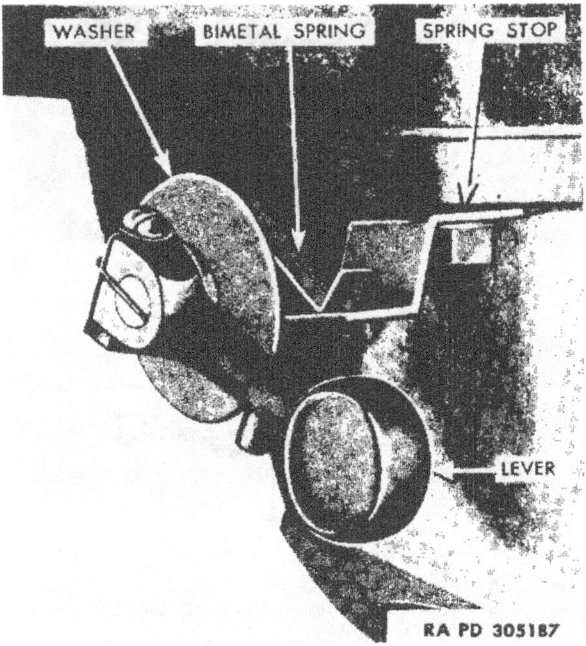

Figure 24—Heat Control Valve

(10) Tighten cylinder head screws and nuts, using torque wrench (par. 54).

(11) Check operation of manifold heat control (par. 53).

52. INTAKE AND EXHAUST MANIFOLDS.

a. Description. The intake and exhaust manifolds (figs. 22 and 23) are attached to each other with four screws, making a unit in which a heat control valve is used to regulate the intake manifold temperature (par. 53).

b. Remove Intake and Exhaust Manifolds. Remove carburetor air horn at top of carburetor, and disconnect hand throttle, choke,

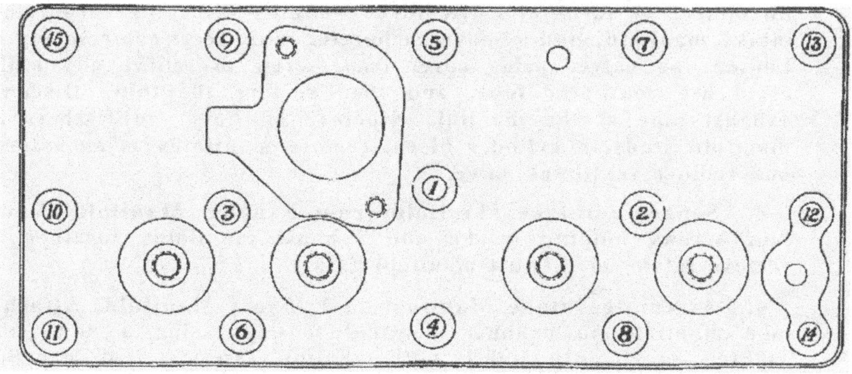

RA PD 305IBS

Figure 25—Cylinder Head Tightening Chart

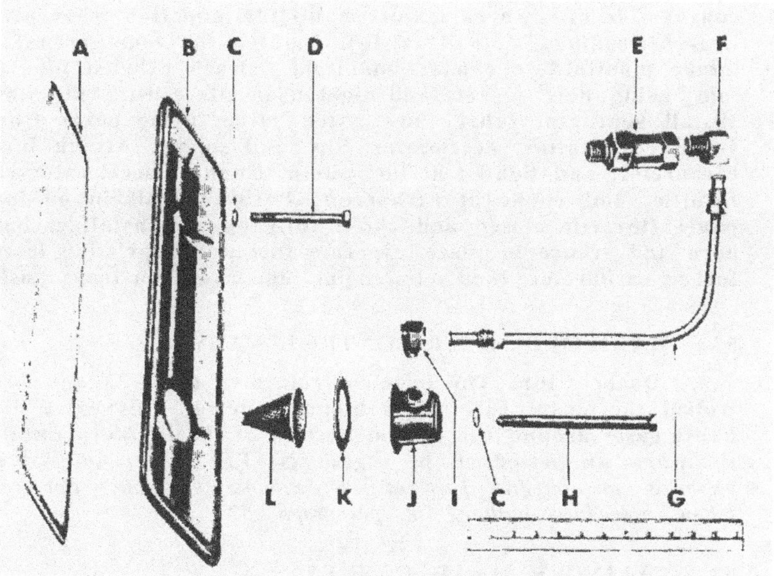

A	VALVE SPRING COVER GASKET	G	CRANKCASE VENTILATOR TUBE
B	VALVE SPRING COVER	H	VALVE SPRING COVER SCREW—FRONT
C	VALVE SPRING COVER SCREW GASKET	I	CRANKCASE VENTILATOR BODY ELBOW
D	VALVE SPRING COVER SCREW—REAR	J	CRANKCASE VENTILATOR BODY
E	CRANKCASE VENTILATOR VALVE	K	CRANKCASE VENTILATOR BODY GASKET
F	CRANKCASE VENTILATOR VALVE ELBOW	L	CRANKCASE VENTILATOR BAFFLE

RA PD 334755

Figure 26— Valve Spring Cover, Disassembled

and accelerator at carburetor. Loosen fuel line at fuel pump, and disconnect at carburetor. Remove two nuts attaching carburetor to intake manifold, and remove carburetor with accelerator spring clip. Loosen the valve spring cover front screw to relieve any pull on crankcase ventilator tube, and then remove the tube. Disconnect exhaust pipe at the manifold. Remove all nuts and washers from manifold studs in cylinder block, remove manifolds as an assembly, and remove ventilator valve.

c. Separate Intake Manifold from Exhaust Manifold. Remove four screws holding intake and exhaust manifolds together, and remove intake to exhaust manifold gasket.

d. Assemble Intake Manifold to Exhaust Manifold. Attach intake manifold to exhaust manifold loosely, using a new gasket. Tighten screws only slightly until manifolds are installed on cylinder block. Install ventilator valve.

e. Install Intake and Exhaust Manifolds. Clean contact surfaces of manifolds and cylinder block. Place new gasket on studs; in cylinder block, and install manifold. Install washers and nuts with convex side of washers against manifolds, and tighten evenly (torque wrench reading 31 to 35 ft-lb). Tighten the four screws attaching intake manifold to exhaust manifold. Attach exhaust pipe to manifold, using new gasket, and tighten in place with nut and screw. Install ventilator tube, and tighten valve spring cover front screw. Install carburetor, accelerator clip, and spring. Attach fuel line at carburetor, and tighten at fuel pump. Connect accelerator rod, hand throttle, and choke at carburetor. Push controls in on instrument panel (throttle closed and choke fully open). Install carburetor air horn and secure in place. Operate fuel pump priming lever to put fuel in carburetor, then start engine, and check for leaky gaskets.

53. MANIFOLD HEAT CONTROL VALVE.

a. Description. The heat control valve (figs. 23 and 24) is controlled thermostatically by a bimetal spring. This valve diverts exhaust gases around the central portion of the intake manifold during the warm-up period of the engine. NOTE: *The manifold heat control valve is an integral part of the exhaust manifold. For replacement follow procedure outlined in paragraph 52.*

54. CYLINDER HEAD GASKET.

a. Removal. Drain the cooling system by opening the drain cock under the radiator at the left front. If there is antifreeze in the cooling system, drain into a pan so it can be used again. Disconnect spark plug wires at the plugs, and remove distributor cap from distributor. Remove two nuts on cylinder studs holding air cleaner tube bracket, and remove bracket with wires and distributor cap. Remove radiator upper tube with hoses attached. Disconnect oil filter upper tube, remove two nuts holding filter to engine, and remove filter. Remove all cylinder head screws and nuts. Remove cylinder head and bonding strap, taking care not to damage oil filler tube, and discard gasket.

b. Installation. Clean cylinder head, tops of pistons, and cylinder block thoroughly. Place cylinder head gasket in position on cylinder block. NOTE: *The front and rear center studs are pilot studs to correctly position the gasket.* Install cylinder head. CAUTION: *Do not damage oil filler pipe.* Install rear bonding strap, oil filter, and air cleaner tube bracket. Install cylinder head bolts and nuts. Tighten cylinder screws and nuts evenly and in sequence (fig. 25), using a

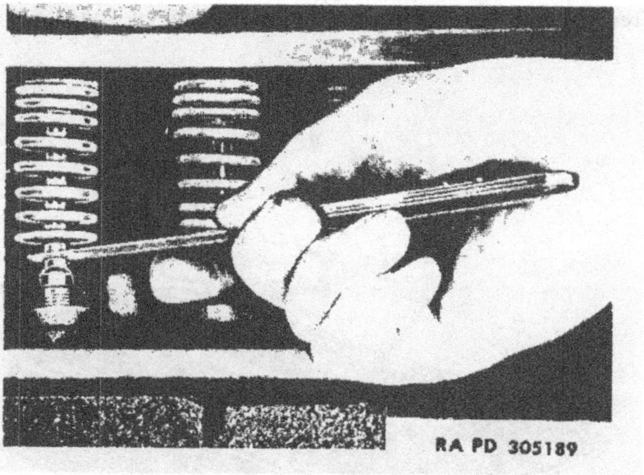

Figure 27—Valve Tappet Adjustment

torque-type wrench (screws, 65 to 70 ft-lb; nuts, 60 to 65 ft-lb). Connect oil filter tube, install distributor cap, and attach sparK plug wires to correct plugs. Install radiator upper tube, tighten hose clamps, and close radiator drain cock. Fill the cooling system, giving due attention to antifreeze, if required (par. 7). Start engine, and check cooling system for leaks. See that cooling solution level has not gone down; replenish if necessary.

55. VALVE COVER GASKET.

a. Removal. Remove front valve spring cover bolt (fig. 26). Remove crankcase ventilator tube at ventilator valve, and remove tube and cap. Remove rear valve spring cover bolt, and slide valve spring cover forward, up, and out over the fuel pump. Discard gasket.

b. Installation. Clean cover and gasket seat on cylinder block. Cement cork gasket to cover. Position cover on cylinder block by sliding it to the rear over fuel pump. Install cover rear screw and copper gasket, but do not tighten. Install cover front screw and copper gasket, with ventilator cap, baffle, and gasket. Connect ventilator tube to valve, and tighten both cover screws evenly. Start engine and check for oil leaks.

56. VALVE TAPPET ADJUSTMENT.

a. **Adjustment.** Remove the valve spring cover (par. 55). Adjust the self-locking tappet screws while they are cold (or warm) to 0.014 inch (fig. 27). Set tappet screws, starting with No. 1 cylinder on compression stroke at top center, then adjust valves in cylinder firing order (par. 62 b), turning the crankshaft one-half turn for each cylinder. NOTE: *The valve tappets will then be on the heel of the cam.* After adjusting, replace valve spring cover (par. 55).

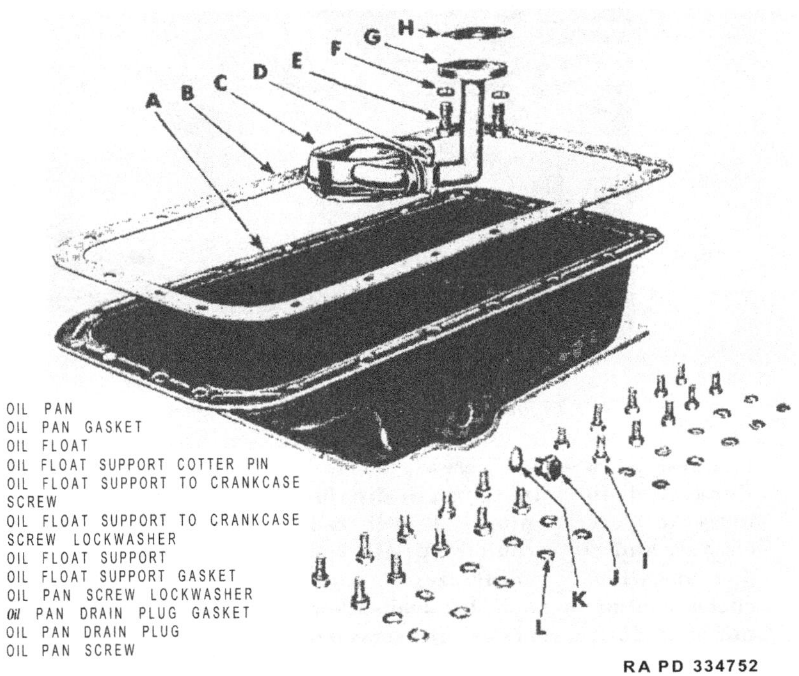

Figure 28—Floating Oil Intake and Oil Pan

57. OIL PAN GASKET.

a. **Removal.** Drain oil by removing drain plug in lower left side of oil pan (fig. 28). Remove oil pan screws, exercising care not to lose spacers under fan belt guard. Remove oil pan, then remove gasket.

b. **Installation.** First clean oil pan thoroughly. Check condition of floating oil intake screen and if dirty, clean in dry-cleaning solution. Clean face of oil pan and crankcase where gasket is installed, and cement gasket to oil pan. Put oil pan in position, and install screws. Be sure that spacers under belt guard are in position, and tighten all screws evenly. Torque wrench reading must be 10 to 14 foot-pounds.

58. OIL FILTER.

a. **Description.** The oil filter is the military standard type located on the right front side of the engine (fig. 36). Part of the oil circulated through the oiling system is sent through the filter. The filter-

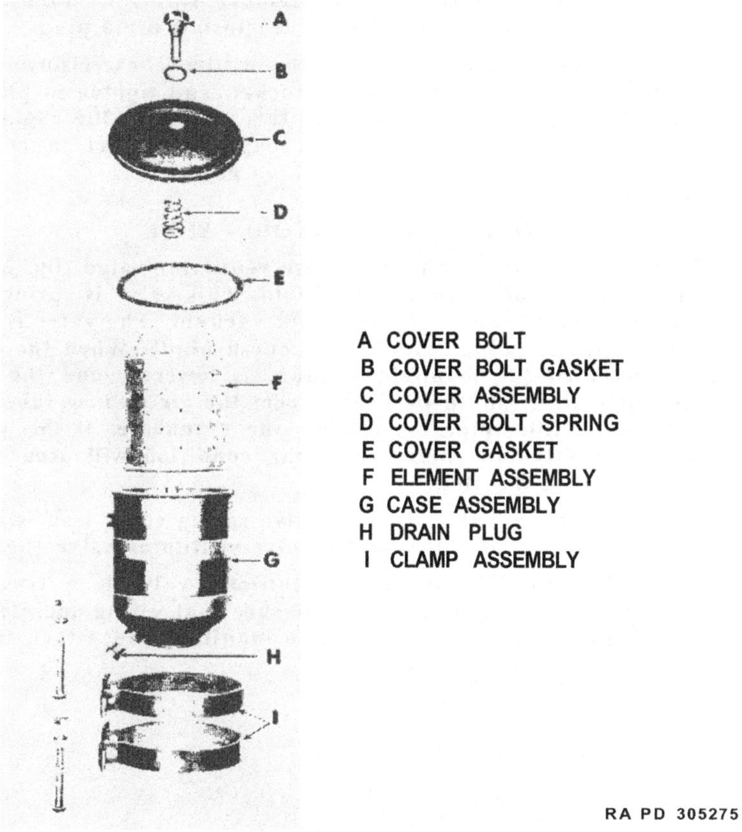

Figure 29—Oil Filter, Disassembled

ing element is a cylindrical replaceable unit which should be changed each 6,000 miles, more often if the oil gets dirty quickly. The inlet line at the top of the filter connects to the oil distribution line at the plug in the left front side of the engine. The outlet or oil return line connects to the timing chain cover.

b. Remove Element. Unscrew cover bolt in top of unit and remove cover, exercising care not to damage gasket (fig. 29). Remove drain plug in lower side of filter to drain filter, then lift out element.

c. Installing Element. Clean filter thoroughly. Install drain plug, and put new element in filter. Inspect cover gasket, and replace if necessary. Install cover and tighten in place with cover bolt. Start engine, and check filter for oil leaks. Add enough oil to crankcase to bring level up to "FULL" mark on gage.

d. Remove Filter. Drain filter by removing drain plug in lower side. Disconnect upper and lower tubes at filter. Remove four bolts holding filter in bracket, and remove filter. Remove tube fittings, using care not to distort them; then install drain plug.

e. Install Filter. Install tube fittings, exercising care not to damage them. Mount filter in bracket, and tighten in place. Connect tubes, being careful not to cross threads. Start the engine and check for oil leaks after which check engine oil level in crankcase and replenish supply to "FULL" mark on gage.

59. CRANKCASE VENTILATOR VALVE.

a. Description. The crankcase ventilator valve (fig. 22) is located at the center of the intake manifold. This valve is spring-loaded, and is operated by the intake manifold vacuum. The valve is closed when the engine is idling (manifold vacuum high). When the engine speed is increased the manifold vacuum is lowered, and the valve opens to allow clean air to be drawn from the air cleaner tube through the engine oil filler pipe to ventilate the crankcase. If this valve fails to seat properly, an engine operating condition will occur similar to a leaky intake manifold.

b. Remove Valve. Loosen valve spring cover front screw. Remove ventilator tube at valve, and remove ventilator valve (fig. 22).

c. Installing Valve. Place ventilator valve in a vise and remove top. Clean the valve and seat. Be sure that spring operates freely, and reassemble valve. Install valve in manifold and attach tube. Tighten valve spring cover front screw.

Section XIV

ENGINE—REMOVAL AND INSTALLATION

60. REMOVAL.

a. Open Hood. Unhook hood by pulling up catches at forward sides of hood. Raise hood and lay back against windshield. Hook or tie hood to windshield to avoid accidental closing.

b. Drain Cooling System. Open drain cocks at lower left-hand corner of radiator, and at right-front lower corner of cylinder block.

c. Remove Battery. Disconnect battery cables. Remove two wing nuts and washers from hold-down bolts. Remove battery hold-down frame, and lift out battery.

d. Remove Radiator. Remove upper and lower radiator hoses. Remove radiator stay rod nuts and remove rod. Remove two radiator stud nuts on bottom of radiator, and lift off radiator. Do not lose radiator pads.

e. Remove Air Cleaner. Disconnect air cleaner flexible hose at cleaner. Loosen wing nuts on side toward engine. Remove nuts on the opposite side, and remove cleaner. CAUTION: *Do not tip and spill oil from reservoir.*

f. Remove Cranking Motor. Disconnect cranking motor wires. Remove two bolts in engine rear plate, and one screw in side of crankcase. Remove motor.

g. On Right Side of Vehicle. Disconnect wires on generator. Disconnect ignition switch wire at ignition coil. Loosen fuel tank cap to relieve any pressure, and disconnect fuel line at flexible connection on right side of engine. Unscrew heat indicator unit from cylinder head. Remove two bolts holding engine front support insulator to frame, and disconnect bond strap.

h. On Left Side of Vehicle. Remove horn from bracket by removing two screws. Remove rear center cylinder head stud nut, and detach bond strap. Disconnect throttle and choke controls at carburetor. Remove fuel line (fuel pump to carburetor). Disconnect oil gage line at upper end of flexible tube on front of dash. Disconnect accelerator rod at lower end of bell crank on back of engine. Remove one bolt and one screw to separate exhaust pipe from manifold. Remove two engine front support bolts in frame, and disconnect engine ground strap from frame.

i. Disconnect Bell Housing. Remove bell housing upper bolts. Wrap a rope or cable around front and rear end of engine, attach chain hoist, and pick up weight of engine. Underneath vehicle, remove the remaining bell housing bolts, and disconnect engine stay cable at the frame crossmember. Drive out two bell housing bolts at side of engine. Raise engine and guide out of frame.

61. INSTALLATION.

a. Installing Engine. Wrap rope or cable around front and rear end of engine, and attach chain hoist. Raise engine and lower into position. Insert transmission shaft in clutch driven plate hub, and work engine back into place. Install dowel bolts from engine side. Install bell housing bolts and tighten. Install engine front support insulator bolts in frame. Attach engine ground strap at left support and bond strap at right support. Install engine stay cable. Run rear adjusting nut up to the bracket, then with cable just taut, tighten lock nut on the front side of bracket.

b. On Right Side of Vehicle. Install cranking motor and attach wires. Attach ignition coil wire. Install heat indicator unit in cylinder head. Attach generator wires and ground strap. Connect flexible fuel line. Check air cleaner oil in reservoir, and install cleaner on dash with wing nuts. Tighten air flexible connection. Install battery, and secure in place with hold-down frame and wing nuts. Clean cable connections, grease, and attach to battery posts.

c. On Left Side of Vehicle. Install accelerator rod with cotter pin. Connect oil gage tube. Attach exhaust pipe to manifold with bolt and screw. Gasket must be in good condition. Install bond strap on cylinder head rear stud, and tighten nut to from 60 to 65 foot-pounds. Install carburetor choke and throttle wires. NOTE: *Controls on instrument panel must be all the way in, the throttle in the carburetor in the closed position, and the choke fully open.* Install fuel line between pump and carburetor. Attach horn to bracket.

d. Installing Radiator. See that radiator pads are in place, and install radiator on frame. Install upper and lower radiator hoses. Install radiator stay rod. Fill radiator, giving due attention to anti-freeze, if required.

e. Inspection. Tighten fuel tank cap. Check engine oil (par. 18 c). Start engine, check for leaks, tune-up, and finally check level of solution in radiator. Close hood and hook properly.

Section XV

IGNITION SYSTEM

Paragraph

62. DESCRIPTION AND DATA.

a. **Description.** The ignition system (fig. 30) is a 6-volt system and consists of the spark plugs, high- and low-tension ignition wires,

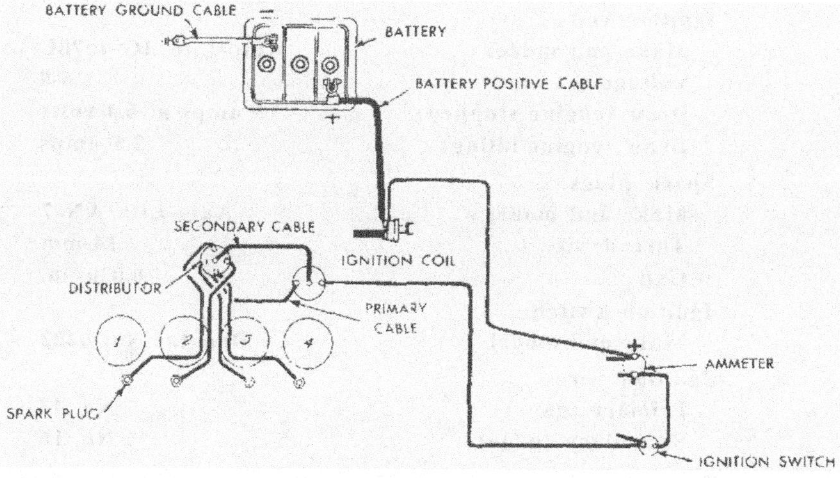

Figure 30—Ignition System Circuit

distributor, coil, and an ignition switch through which it is connected to the electrical system of the vehicle. There are two separate circuits in the ignition system (primary and secondary) which combine to develop the high-voltage current necessary to make a spark jump the plug gaps in the engine combustion chambers, and ignite the fuel mixture. In operation, with the ignition switch turned on, and the distributor points closed, current flows through the primary winding of the ignition coil and builds up a strong magnetic field. When the distributor points open, the magnetic field collapses and induces a high-voltage current in the secondary winding of the coil. This happens each time the points open. This high-voltage current is delivered to the spark plugs at the correct time by the distributor rotor, cap, and secondary wires. To prevent burning of the distributor points by current arcing across the open points, a condenser is connected across the points (in parallel). This provides a place (capacity) for current to go until the points open enough to prevent arcing. Discharge of this condenser current back through the primary winding of the coil causes the magnetic field to collapse much faster in developing the high-voltage current for the spark plugs.

b. Data.

Distributor

Make and model	Auto-Lite IAD-4008
Type advance	Centrifugal
Rotation	Counterclockwise
Firing order	1-3-4-2
Point gap	0.020 in.
Breaker arm spring tension	17 to 20 oz
Condenser capacity	0.18 to 0.26 mfd

Ignition coil

Make and model	Auto-Lite IG-4070L
Voltage	6-8
Draw (engine stopped)	5 amps at 6.4 volts
Draw (engine idling)	2.5 amps

Spark plugs

Make and model	Auto-Lite AN-7
Thread size	14-mm
Gap	0.030 in.

Ignition switch

Make and model	Douglas No. 6282

Ignition wires

Primary (gage)	No. 14
Secondary (gage)	No. 16

c. Tests. The following procedure will assist in localizing trouble in the ignition system without the use of instruments:

(1) First check the brilliancy of the headlights and operate cranking motor, to judge the condition of the battery and connections as far as the ammeter.

(2) Remove an ignition wire at a spark plug and hold about three-eighths inch away from a bare metal part of the engine. A good spark should result when the cranking motor is operated with the ignition switch on. If spark is weak or absent, proceed as follows:

(3) Pull coil wire out of center of distributor cap; remove cap and crank engine until distributor points are fully closed. Turn on ignition switch, hold cap end of coil secondary wire about three-eighths inch away from cylinder block, and open breaker points with

the fingers, or rock the distributor cam. If a good spark occurs, fault is located in distributor cap, rotor, or wires; inspect cap and rotor for cracks or carbon runners and ignition wires for short circuits. To test distributor cap place wire back in cap and operate cranking motor. Short circuit will be evidenced by a spark within or outside the cap. To check rotor, remove from distributor; put end of coil secondary wire in rotor, and hold top against cylinder block; operate distributor points (spark will show where "short" occurs). If no spark occurs in preceding test, proceed as follows:

(4) Open distributor points and notice if a slight spark is obtained. If spark occurs, current is reaching points. If no spark occurs, detach condenser, and repeat above operation. If spark is obtained, condenser is at fault, and must be replaced. If no spark is obtained, determine if coil primary wires are faulty, as follows:

(5) Remove switch wire at ignition coil, and strike wire terminal against cylinder block. If no spark is obtained, check wiring up to ammeter for loose connection or open circuit. If spark is obtained, current is reaching coil, and this indicates coil is faulty. Replace coil.

63. MAINTENANCE.

a. The distributor requires periodic lubrication at various points (par. 18). Keep coil and distributor wires pushed down in towers. Keep distributor cap and spark plug porcelains free from dirt and grease. All wire terminals must be clean and tight. Replace any wires that are frayed or have cracked insulation. Clean and adjust spark plugs and distributor points (par. 67).

64. DISTRIBUTOR.

a. Description. The distributor (fig. 32) is mounted on the right side of the engine. A full automatic spark advance is mechanically governed by two counterweights which advance the spark as the engine speed increases. The distributor is driven by a shaft extending into the oil pump driven gear, which is driven by a gear on the camshaft. The lower end of the distributor shaft has an offset tongue which must be in correct relation to the oil pump shaft before the two can be assembled. A friction spring on the distributor shaft (fig. 21) engages in the oil pump gear to prevent backlash at this point, and uneven engine performance.

b. Removal. After raising hood, remove wires from distributor cap and primary wire from terminal on side of distributor. Remove screw holding advance arm to crankcase, and pull out distributor assembly.

c. Installation. Remove distributor cap. Insert distributor in crankcase, pushing it down into place, turn rotor until offset tongue on lower end of distributor shaft fits into oil pump shaft, then push farther down into position. Some resistance will be experienced, caused by friction of the spring on the lower end of the distributor shaft fitting into the oil pump gear (fig. 21). Install screw in advance arm loosely. Attach primary wire to distributor. Set timing (par. 65). Install distributor cap and wires (fig. 32).

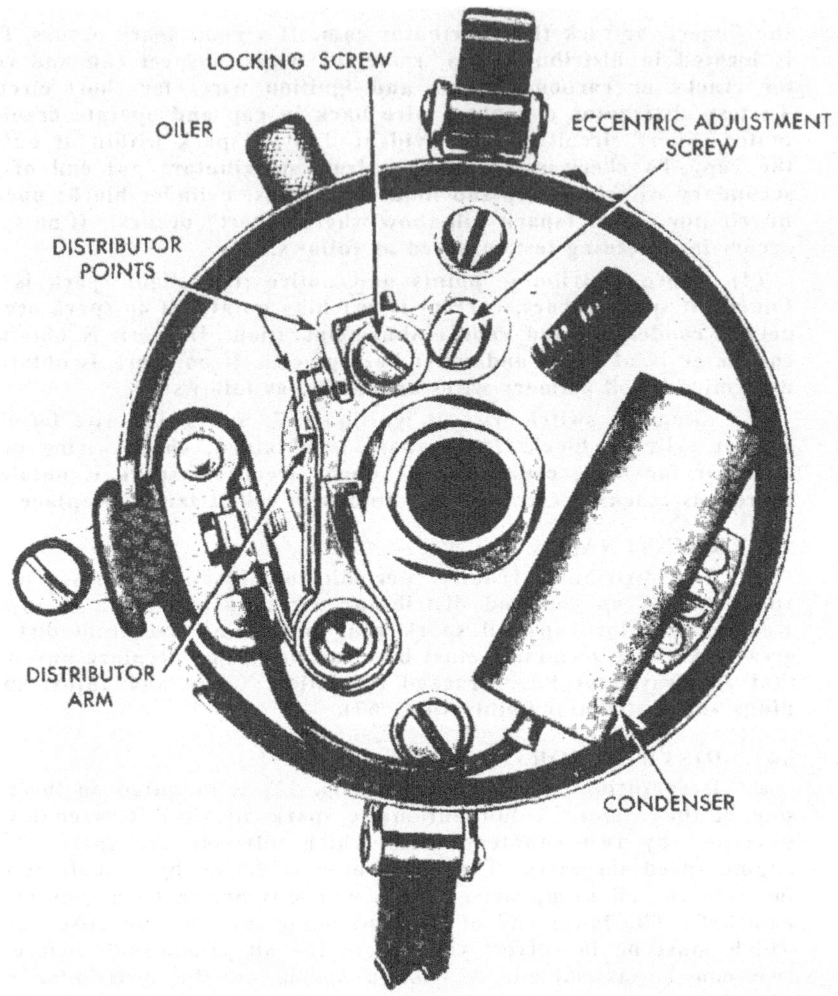

Figure 31—Distributor Points and Condenser

d. **Distributor Points.**

(1) ADJUSTMENT. Slip off the two clips holding the distributor cap in place and remove cap. Lift off rotor. Crank engine until point arm rubbing block is on top of a cam. Loosen lock screw (fig. 31), and turn eccentric screw until point gap is 0.020 inch measured with a thickness gage. Tighten lock screw and recheck gap. Install rotor and cap. Push wires well down into cap.

(2) REMOVAL. Slip off the two clips holding the distributor cap in place and lift off cap, then remove rotor (fig. 31). Using a small screwdriver, unscrew condenser lead which will release breaker arm

RA PD 305 1 95

Figure 32—Distributor Wires

spring, then lift off breaker arm. Remove screw in stationary breaker point, and lift out point.

(3) INSTALLATION. Place stationary breaker point in distributor, and install locking screw loosely. Lightly lubricate breaker arm pivot pin, and install breaker arm. Place spring in position with condenser lead, insert screw, and tighten securely in place. Aline points if necessary. To adjust points refer to step (1) above.

e. Distributor Condenser.

(1) DESCRIPTION. The condenser is attached by one screw to the support plate in the distributor, and connected by a short flexible wire across the distributor points. The condenser functions to absorb momentarily any current which has a tendency to arc across the points when they open. The condenser must be firmly attached to the

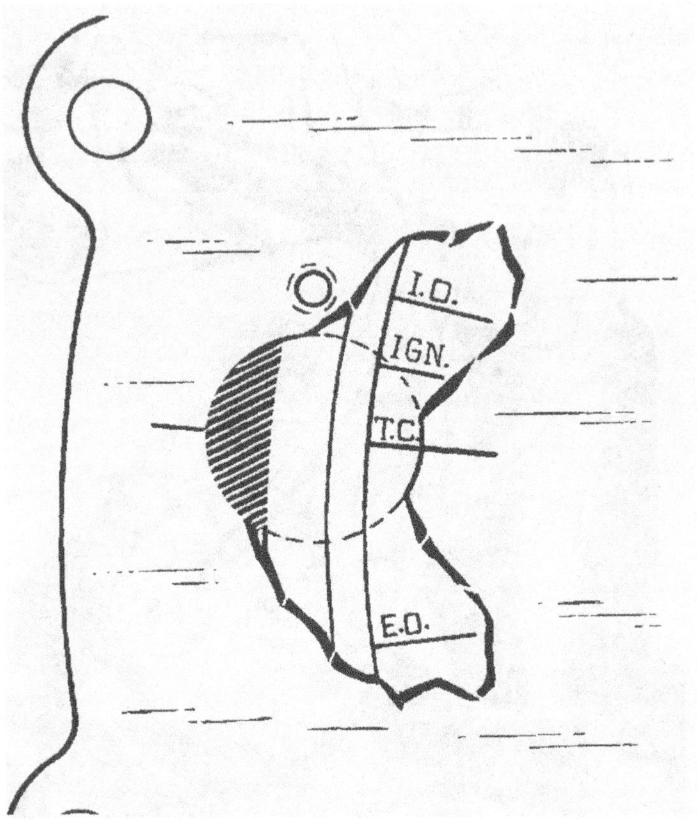

RA PD 305193

Figure 33—Timing Marks (Flywheel)

support plate, and the cable in good condition. Refer to paragraph 62 c for tests. Test condenser on a condenser tester, if available.

(2) REMOVAL. Slip off distributor cap clips and remove cap. Lift off rotor. Remove screw holding condenser to support plate. Remove screw in lead and remove condenser.

(3) INSTALLATION. Position condenser on plate and attach with screw. Attach lead. NOTE: *Determine if distributor points have been disturbed.* Install rotor and cap.

65. IGNITION TIMING.

a. To Set Timing without Timing Light. Remove timing hole cover on engine rear plate at right side under cranking motor. Remove distributor cap. Crank engine to No. 1 cylinder firing stroke (distributor rotor toward lower front corner of distributor). Set flywheel with ignition mark (fig. 33) in center of timing hole. Turn distributor housing so that breaker points are just opening, and tighten screw holding distributor to engine. Install timing hole cover.

b. To Set Timing with Timing Light. Remove timing hole covet on engine rear plate at right side under cranking motor. Attach one lead of timing light to No. 1 spark plug, without removing the high-tension wire, and ground the other lead to the engine. Start engine and run slowly. Hold timing light in position to illuminate timing hole. Observe timing mark on flywheel, as illuminated by light, in relation to timing mark on engine plate. If marks do not coincide, loosen distributor clamp screw, and turn distributor housing in proper direction until marks do coincide. NOTE: *Use mirror for a better view.* Tighten clamp screw. Accelerate engine speed, and with light observe timing marks. They should separate to indicate that centrifugal spark advance is functioning.

66. IGNITION COIL.

a. Description. The ignition coil is mounted on the right side of the engine at the rear (fig. 21). A terminal is provided for the primary wire from the switch, a terminal for the primary wire to the distributor, and a terminal for the high-tension wire to the distributor cap, and a ground strap connection. The coil steps up the primary current, furnished by the battery and generator, to high-tension current of sufficient voltage to cause a spark to jump the gaps of the spark plugs.

b. Removal. Remove air cleaner by loosening clamp on flexible tube to carburetor; loosen wing nuts on air cleaner bracket at center of dash, and remove those on the right side. Pull secondary wire out of top of coil. Disconnect primary wires, then remove coil and bracket from engine, after which remove the bond strap.

c. Installation. Attach bond strap to coil, and mount coil on engine with bond strap on coil bracket front stud, and tighten securely. Attach primary wires, and push secondary wire into top of coil. Install air cleaner, and tighten hose clamp.

67. SPARK PLUGS.

a. Description. Spark plugs (fig. 22) are located in the top of the cylinder head at the left side. The plugs are of the one-piece type. A copper-silver alloy gasket is used on each plug for heat transfer to the cylinder head, and to prevent leakage of compression. Push-on type wire terminals are used with radio filters at each plug. A spark plug insulator cap is fitted on each plug to protect the plugs from water and dirt. No shielding is used on the plugs.

b. Adjustment. To adjust the gap, bend the side electrode only, and gage the plug with a round thickness gage to a gap of 0.030 inch.

c. Removal. To avoid breakage of the porcelain, remove the plugs with the socket wrench and handle furnished in the vehicle tool equipment.

d. Installation. Install new plug gaskets if available. Tighten spark plugs snug so that gasket will compress.

68. IGNITION SWITCH.

a. Description. On the earlier production, a key-type ignition switch was used which has been superseded by an interchangeable lever-type switch (fig. 5). Turn the switch lever clockwise for "ON" position.

b Removal. Disconnect battery positive cable at battery. Unscrew retaining nut against face of instrument panel, and remove switch from panel. Disconnect wires and remove switch.

c. Installation. Install the wires on proper terminals, then fit the switch into the hole in the instrument panel, screw on retaining nut, and tighten securely.

69. IGNITION WIRING.

a. Description. The ignition wiring (fig. 30) consists of low-tension and high-tension wires. The low-tension, or primary wires, carry the current from the ignition switch, which is connected to the electrical system of the vehicle, to the ignition coil, and from the coil to the distributor. The high-tension or secondary wires carry the high-voltage current generated in the secondary wires of the coil, to the distributor, where it is distributed to the spark plugs. No shielding harness is used.

b. Removal. Before removing ignition switch wire, disconnect the battery negative (ground) cable. It is not necessary to disconnect the battery cable to replace the coil-to-distributor primary wire or the secondary wires. Disconnect wires at terminals, and remove wire or harness, opening wire clips on the harness in which the switch wire is a part. When removing secondary wires mark terminal tower in distributor cap for No. 1 cylinder spark plug. Pull wires out of distributor cap and off spark plug terminals.

c. Installation. To install primary wires, run wire or harness through clips, and attach terminals securely. Install secondary wires, for spark plugs, through support bracket. Push terminals down well into the proper distributor towers (fig. 32). Push rubber secondary wire tip down in place on distributor towers. Push terminals down on proper spark plugs. Refer to firing order (par. 62 b).

Section XVI

FUEL AND AIR INTAKE AND EXHAUST SYSTEMS

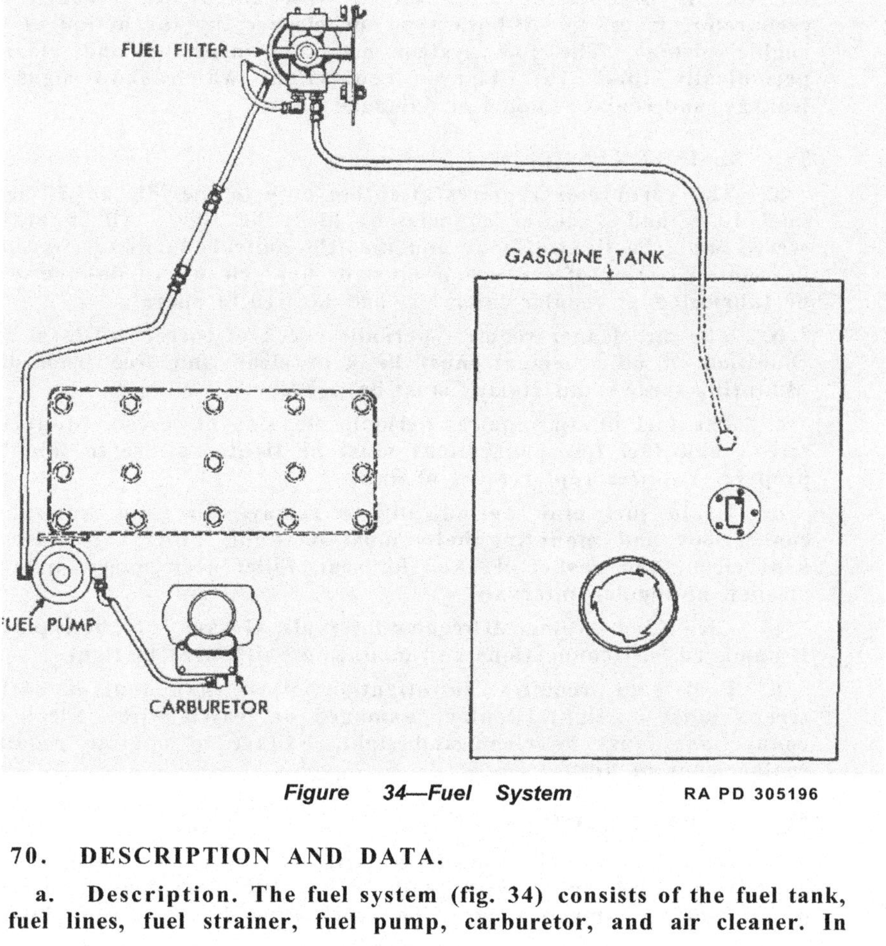

Figure 34—Fuel System RA PD 305196

70. DESCRIPTION AND DATA.

a. **Description.** The fuel system (fig. 34) consists of the fuel tank, fuel lines, fuel strainer, fuel pump, carburetor, and air cleaner. In addition to these units an electric-type fuel gage is mounted on the instrument panel, and is connected by one wire to a fuel tank unit.

b. Data.

Carburetor	Carter WO-539S
Air cleaner	Oakes 613300
Fuel pump	AC 1538312
Fuel pump static pressure	4.5 lb at 1,800 rpm
Fuel tank capacity	15 gal
Fuel strainer	Type T-2; AC1595848
Fuel gage	Electric actuation

c. Operation. Fuel in the tank is drawn through the fuel strainer by the action of a pump mounted on the forward left side of the engine. The pump also forces the fuel into the carburetor bowl until the flow is shut off by the carburetor float valve. In the carburetor, the fuel is proportioned and mixed with air drawn through the carburetor, from the oil-bath type air cleaner, by the action of the engine pistons. The fuel system must be inspected and cleaned periodically (par. 16). Tighten connections which show signs of leakage, and replace kinked or damaged lines.

71. MAINTENANCE.

a. The carburetor requires attention only to the idle adjustment. Fuel lines and vacuum connections must be tight. All mounting screws must be tight. Choke and throttle control clamp screws must be tight. Exterior of carburetor must be kept clean. All linkage must be lubricated at regular intervals, and be free to operate.

b. The air cleaner requires periodic check of correct oil level and condition of oil. Element must be kept clean and free from dirt. Mounting screws and clamps must be tight.

c. The fuel pump requires periodic cleaning of screen. Mounting screws and fuel line connections must be tight. Failure to function properly requires replacement of unit.

d. Drain fuel tank periodically to remove dirt and water. All connections and mounting bolts must be tight. Filler cap must be kept clean, and gasket checked for seat. Filler neck screen must be cleaned at regular intervals.

e. Clean fuel strainer at regular intervals. Gasket must be replaced if damaged. All connections and mounting bolts must be tight.

f. Fuel gage requires no attention other than that mounting screws must be tight. Replace damaged or frayed wires. Electrical connections must be clean and tight. Failure to operate requires replacement of unit.

72. CARBURETOR.

a. Description. The carburetor (fig. 35) is of the conventional downdraft, plain-tube type, with a throttle operated accelerator pump and economizer device. The carburetor is a precision instru-

Figure 35—Carburetor Idle Adjustment

ment which delivers the proper fuel and air mixture for all speeds and operating requirements of the engine.

b. Adjustment. The idle adjustment screw indicated in figure 35 is the only service adjustment provided on the carburetor. To obtain the approximate correct setting, turn the adjustment screw to the right and all the way in, but do not jam the screw against the seat, then, back out adjustment screw between one and two turns. To make the final adjustment, warm up the engine, and adjust the screw until the engine runs smoothly. Set the throttle stop screw so the engine will idle at 600 revolutions per minute (vehicle speed, 8 mph). Replace carburetor if it requires other attention.

c. Removal. Loosen clamp on air horn and flexible air hose, and remove air horn. Remove throttle and choke control wires. Disconnect throttle control rod at throttle lever. Disconnect fuel line at carburetor. Remove carburetor flange nuts and retracting spring clip, and lift off carburetor.

d. Installation. Inspect condition of gaskets between carburetor and manifold, and install new gaskets, if required. Install carburetor retracting spring clip, and carburetor flange nuts. Tighten nuts evenly. Connect throttle control rod, and throttle and choke wires. Install carburetor air horn, and tighten clamp screws. Adjust carburetor (subpar. b above).

Figure 36—Air Cleaner and Oil Filter

73. AIR CLEANER.

a. Description. The air cleaner (fig. 36) is of the oil-bath type, and mounted on the right-front side of the dash. Air enters through louvers in the dash side of the unit, passes down and across the surface of the oil, up through the filtering element, then to the carburetor. For servicing of air cleaner refer to subparagraph d below.

b. Removal of Oil Cup. Hold one hand under cup and spring loose the two retaining clamps. Remove cup, clean, and refill to indicated oil level.

c. Installation. Place cup in position on bottom of cleaner, push up into place, and lock with the retaining clamps.

d. Removal and Servicing of Air Cleaner. Loosen the hose clamp and two wing nuts at the center of the dash. Remove two wing nuts on right side of dash, and lift out cleaner assembly. Unfasten the two clamps holding oil cup in place, and remove cup (fig. 37). Unscrew element wing bolt in bottom center of cleaner, and pull out cleaner element. Plunge the element up and down in dry-cleaning solvent to wash out any dirt, then dry with compressed air.

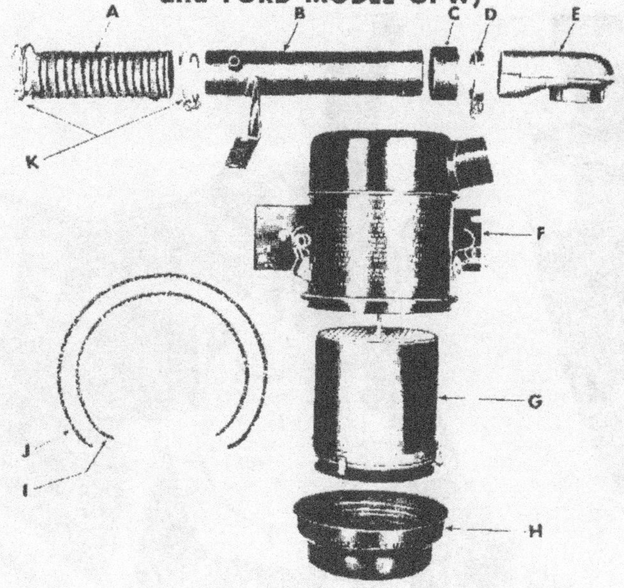

A FLEXIBLE HOSE CONNECTION
B TUBE AND BRACKET ASSEMBLY
C BUSHING
D HOSE CLAMP
E CARBURETOR AIR CLEANER HORN
F BODY ASSEMBLY
G ELEMENT AND WING BOLT ASSEMBLY
H CUP
I LOWER GASKET
J BODY GASKET
K HOSE CLAMP

RA PD 305281

Figure 37—Air Cleaner, Disassembled

e. **Installation.** Install the element in cleaner housing, and secure in place with retaining bolt. Clean and refill oil cup to indicated level, and clamp into place on cleaner. Mount cleaner on dash, install and tighten mounting nuts, and tighten air hose clamp in place.

74. FUEL PUMP.

a. **Description.** The fuel pump (fig. 38) located on the forward, left side of the engine is a diaphragm type, operated by a lever against an eccentric on the engine camshaft. The pump is equipped with a hand lever on the rear side which can be used to operate the pump for priming the carburetor bowl. Lever must be placed down for camshaft to operate pump. A filter screen in incorporated in the fuel bowl, and should be cleaned periodically (subpar. b below).

b. **To Clean Screen.** Unscrew knurled nut on bowl clamp, swing clamp aside and lift off bowl. Clean screen with dry-cleaning solvent and small brush. Dry with compressed air. Blow out fuel chamber lightly with compressed air and install screen in place. Install new bowl gasket or turn over old one if serviceable. Install bowl, place bail clamp in position and tighten securely.

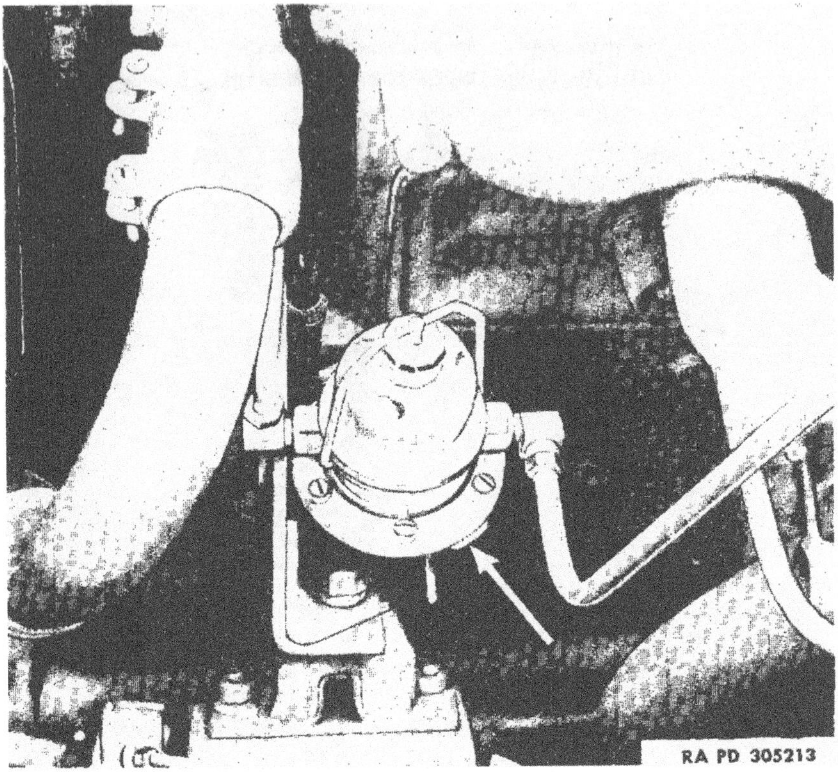

Figure 38—Fuel Pump

c. Removal. Remove the inlet and outlet lines, remove two screws holding pump to side of engine and remove pump.

d. Installation. Install pump in place on crankcase and tighten securely with two screws. Inspect gasket and replace if unserviceable. Attach inlet and outlet lines and tighten. Prime pump by operating priming lever, start engine, and check connections for leaks.

75. FUEL TANK.

a. Description. The fuel tank (fig. 39) is located under the driver's seat. An extension filler neck can be pulled up to facilitate filling the tank from a container. After removing the filler cap, pull up on the filler extension, and turn to the right to lock it in place. To remove the filler extension from the tank turn it to the left and pull up.

b. Removal. Drain fuel by removing drain plug in left side of tank. Remove bolts in seat rear flange and front legs, then lift out seat. Remove filler cap. Disconnect fuel gage wire and remove fuel

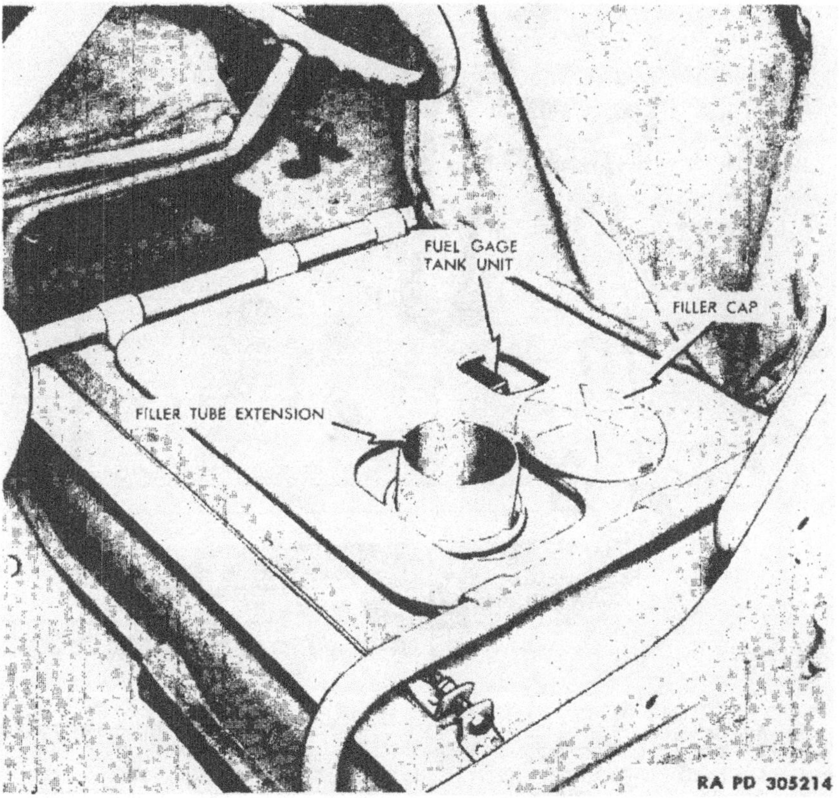

Figure 39—Fuel Tank

gage unit by taking out five screws. Disconnect fuel line from tank, under vehicle. Remove bolt holding tank to wheel housing, remove bolts in tank straps, and lift out tank.

c. Installation. Clean out fuel tank sump in body, and install tank in place. Attach tank straps and bolts, also bolt holding tank to wheel housing. Connect fuel line to tank. Install fuel gage in tank, and attach wire to gage. Set seat in position and secure in place with bolts in front legs and seat flange. Fill tank, install cap and check tank and connections for leaks.

76. FUEL STRAINER.

a. Description. The fuel strainer (fig. 41) has a disk-type (laminated) element, and a settling bowl for dirt and water. It is located between the fuel tank and the pump, and is mounted on the right-front side of the dash. Service this unit in accordance with preventive maintenance procedure (par. 16).

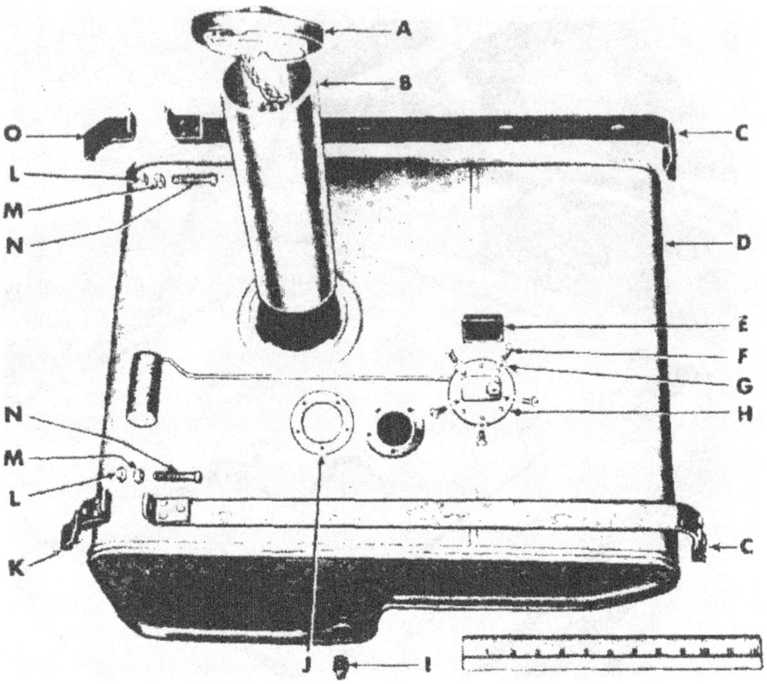

A CAP
B FILLER TUBE EXTENSION
C HOLD DOWN STRAP
D FUEL TANK ASSEMBLY
E GAGE TANK UNIT PROTECTOR
F GAGE TO TANK SCREW
G GAGE TO TANK SCREW LOCKWASHER
H GAGE TANK UNIT ASSEMBLY
1 DRAIN PLUG
J GAGE GASKET
K HOLD DOWN CLAMP
L HOLD DOWN STRAP LOCK NUT
M HOLD DOWN STRAP NUT
N STRAP CLAMP SCREW
O STRAP CLAMP

RA PD 334756

Figure 40—Fuel Tank, Disassembled

b. To Clean Strainer. Remove drain plug and allow strainer to drain into a container. NOTE: *Do not allow fuel to drain onto cranking motor.* Unscrew cover bolt and remove strainer bowl and element. Do not damage bowl gasket. Remove element from bowl, and clean thoroughly in dry-cleaning solvent. Be sure all dirt particles are removed from between disks of the element. Blow element dry with compressed air, but do not use extreme pressure. Wash strainer bowl, and dry with clean cloth. Install element spring and element in bowl. Check condition of element gasket, bowl gasket and cover bolt gasket, and replace, if damaged. Install gaskets in position, and assemble bowl to cover. Install cover bolt and gasket, and tighten securely. Install drain plug. Start engine, run a few minutes, stop engine, and check for leaks.

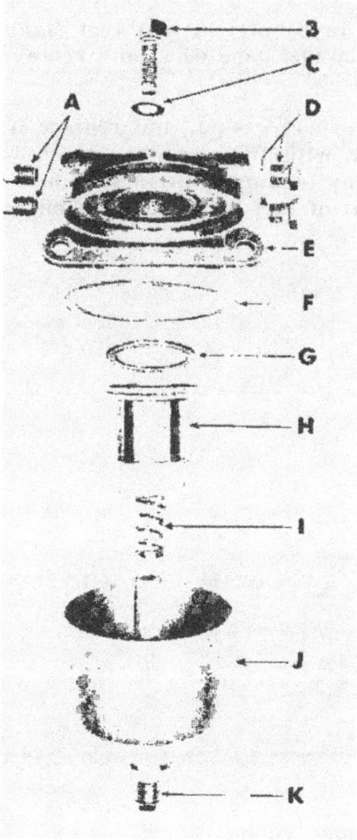

A PIPE PLUG
B COVER CAP SCREW
C COVER CAP SCREW GASKET
D PIPE REDUCING BUSHING
E COVER
F STRAINER BOWL GASKET
G STRAINER UNIT GASKET
H STRAINER UNIT ASSEMBLY
I STRAINER UNIT SPRING
J STRAINER BOWL AND CENTER STUD ASSEMBLY
K STRAINER DRAIN PLUG

RA PD 305282

Figure 41—Fuel Strainer, Disassembled

c. **Removal.** Disconnect inlet and outlet lines from strainer (fig. 36). Open glove compartment door, and remove mounting bolt nuts, then remove strainer.

d. **Installation.** Place mounting bolts in strainer bracket, and reinstall on dash. Install nuts on bolts through glove compartment. Attach inlet and outlet lines to strainer. Start engine, run a few minutes, stop engine, and check for leaks.

77. FUEL GAGE.

a. **Description.** The fuel gage consists of an electrical indicating unit located in the instrument panel, and a tank unit in the fuel tank, for actuating the instrument panel unit. A float in the tank operates a rheostat, which governs the current flowing through the instrument panel gage. The gage registers only while the ignition switch is on. The fuel gage wiring circuit is shown in figure 42.

b. Removal of Tank Unit. Remove bolts in seat rear flange and front legs. Take out seat. Disconnect fuel gage wire, and remove tank unit, by unscrewing five screws.

c. Installation of Tank Unit. Inspect gasket, and replace if damaged. Install tank unit, and secure with five screws. Attach wire to fuel gage. Check operation of gage by turning ignition switch to "ON" position; gage must register amount of fuel in tank. Turn off ignition

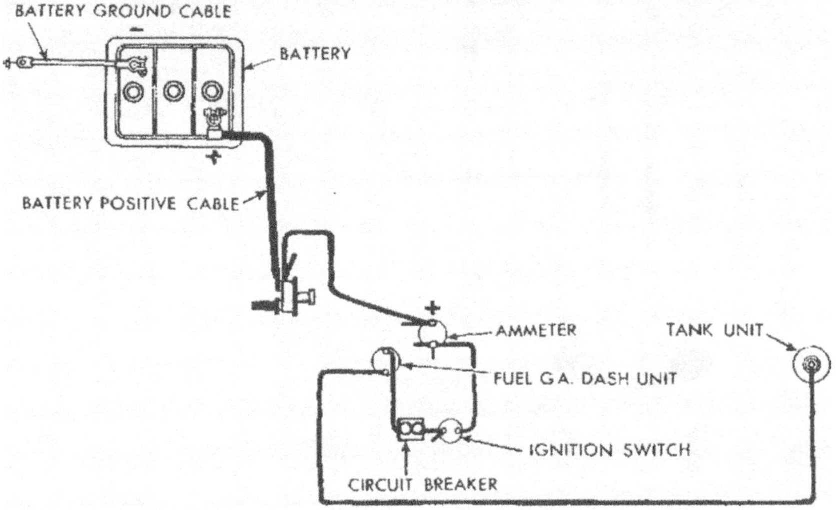

RA PD 305198

Figure 42—Fuel Gage Wiring Circuit

switch. Install seat, and secure in place with bolts in front legs and seat rear flange.

d. Removal of Fuel Gage (Instrument Panel). Disconnect positive terminal cable at battery. Remove wires on back of gage, and two retaining nuts on the retaining bracket. Remove bracket and take gage out through front of panel.

e. Installation of Fuel Gage (Instrument Panel). Place gage in panel, and install retaining clamp and nuts. Position gage correctly, and tighten retaining clamp nuts. Attach wires (wire to tank unit is attached to left terminal; wire from circuit breaker attaches to right-hand terminal). Clean terminal, and install positive cable on battery.

78. EXHAUST SYSTEM.

a. Description. The exhaust system (fig. 43) consists of the exhaust pipe which, for maximum road clearance, passes under the vehicle to the muffler located under the right side of the body. A flexible section of the exhaust pipe permits movement of the engine in the frame. The muffler is mounted on brackets with fabric inserts.

b. **Removal of Exhaust Pipe.** Remove three bolts in exhaust pipe shield at right frame member and remove shield. Remove two exhaust pipe clamp bolts in skid plate under transmission. Loosen clamp on exhaust pipe, at front end of muffler. Remove bolt and

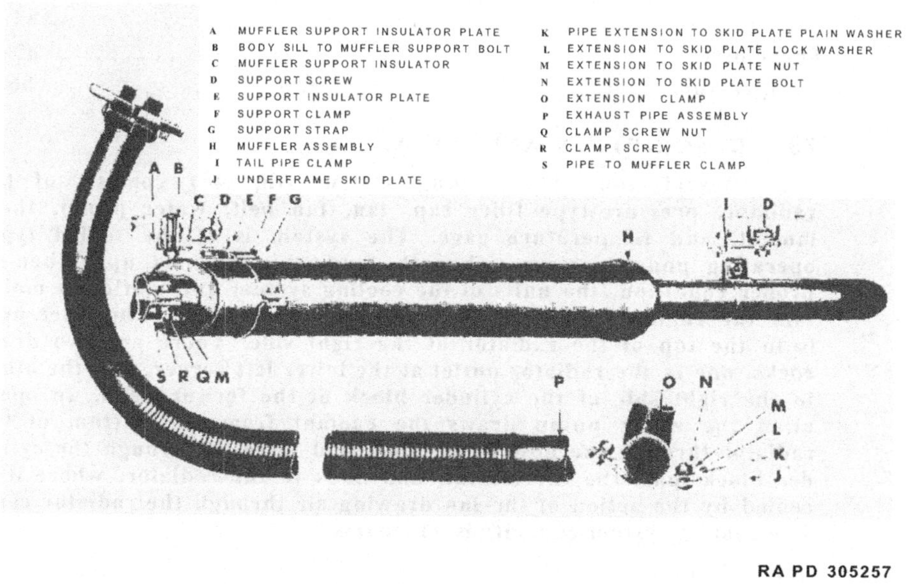

Figure 43—Exhaust System

screw in exhaust pipe flange at exhaust manifold. Drop front end of exhaust pipe, and remove from the left side of vehicle. Discard gasket.

c. **Installation of Exhaust Pipe.** Install pipe in position, and insert rear end in muffler. Install exhaust pipe manifold flange gasket, and attach pipe to manifold with screw and nut, and tighten evenly. Attach clamp at muffler, and attach exhaust pipe shield with three bolts.

d. **Removal of Muffler.** To remove the muffler, loosen exhaust pipe clamp. Remove muffler front support bolt and tail pipe support bolt. Remove muffler assembly, after which, remove muffler strap and tail pipe clamp.

e. **Installation of Muffler.** Loosely install tail pipe clamp and support clamp on muffler. Place muffler on exhaust pipe. Install bolt through tail pipe clamp and flexible mounting. Tighten tail pipe clamp bolt. Install bolt in muffler support and flexible mounting. Tighten muffler strap bolt and exhaust pipe clamp.

Section XVII

COOLING SYSTEM

79. DESCRIPTION AND DATA.

a. Description. The cooling system (fig. 44) consists of the radiator, pressure-type filler cap, fan, fan belt, water pump, thermostat, and temperature gage. The system is of the sealed type, operating under pressure when the engine is warmed up. When in proper condition, the units of the cooling system automatically maintain the engine at the proper operating temperature. The filler pipe is in the top of the radiator at the right side. There are two drain cocks, one in the radiator outlet at the lower left corner, and the other in the right side of the cylinder block at the forward end. In operation the water pump draws the coolant from the bottom of the radiator through the hose connection, and forces it through the cylinder block, past the thermostat, and back to the radiator, where it is cooled by the action of the fan drawing air through the radiator core. The cooling system capacity is 11 quarts.

b. Data.

Cooling system
- Capacity .... 11 qt

Radiator
- Type .... Fin and tube
- Filler cap .... Pressure type

Water pump
- Type .... Centrifugal
- Drive .... Fan belt
- Bearings .... Prelubricated ball

Fan belt
- Type .... Vee
- Length .... 44⅛ in.
- Width .... $1\frac{1}{16}$ in.
- Angle of Vee .... 42 deg

Fan
- Blades 4
- Diameter .... 15 in.

Thermostat
  Opens . . . . 145° to 155°F
  Fully open . . . . 170°F
Temperature gage type . . . . Capillary

80. MAINTENANCE.

a. The cooling system must be inspected in accordance with preventive maintenance procedures (pars. 13 and 16). When draining the cooling system refer to caution plate on the instrument panel (fig. 7). General maintenance of the cooling system consists of the following procedures.

(1) Keep sufficient coolant in the system. Use clean water to which must be added the specified rust inhibitor; at temperatures below 32°F, add proper quantity of antifreeze solution (par. 7).

(2) Drain, flush, and refill system whenever inspection reveals any accumulation of rust or scale. Clean system seasonally as well as before and after using antifreeze solution.

(3) If engine overheats due to lack of coolant in the system, do not add cold water immediately. Let engine cool so that radiator does not boil, start engine, and add water slowly to prevent damage to cylinder block and head.

(4) Do not overfill radiator. Fill radiator to bottom of baffle visible through the radiator filler hole.

(5) Keep cylinder head, water pump, hose clamps, and connections leakproof. Replace deteriorated or leaky hose.

(6) Adjust fan belt and replace as required.

(7) Test periodically for air suction and exhaust gas leaking into system (subpars. c and d below).

b. Draining and Refilling System.

(1) Drain the cooling system, when required, by opening the drain cocks at the lower left corner of the radiator, and at the right front corner of the cylinder block. Loosen the pressure-type radiator cap to break any vacuum which might prevent proper draining. If solution is to be saved, catch it in a clean container. If system is not to be refilled immediately, attach a tag to the steering wheel, warning personnel about the system being drained.

(2) Refill the cooling system by first closing the two drain cocks tightly. Use clean water available, preferably soft water (water with low alkali content or other substances that promote rust and scale). Fill system through radiator filler pipe until level is up to lower edge of baffle visible through filler hole. Install radiator cap, and turn clockwise to tighten. Start engine and warm up. Check coolant level in radiator, and add more if required.

c. Air Suction Test. The air suction test is used to determine if air is entering the coolant, possibly due to low coolant level in the radiator, leaky water pump, or loose hose connections. To make test,

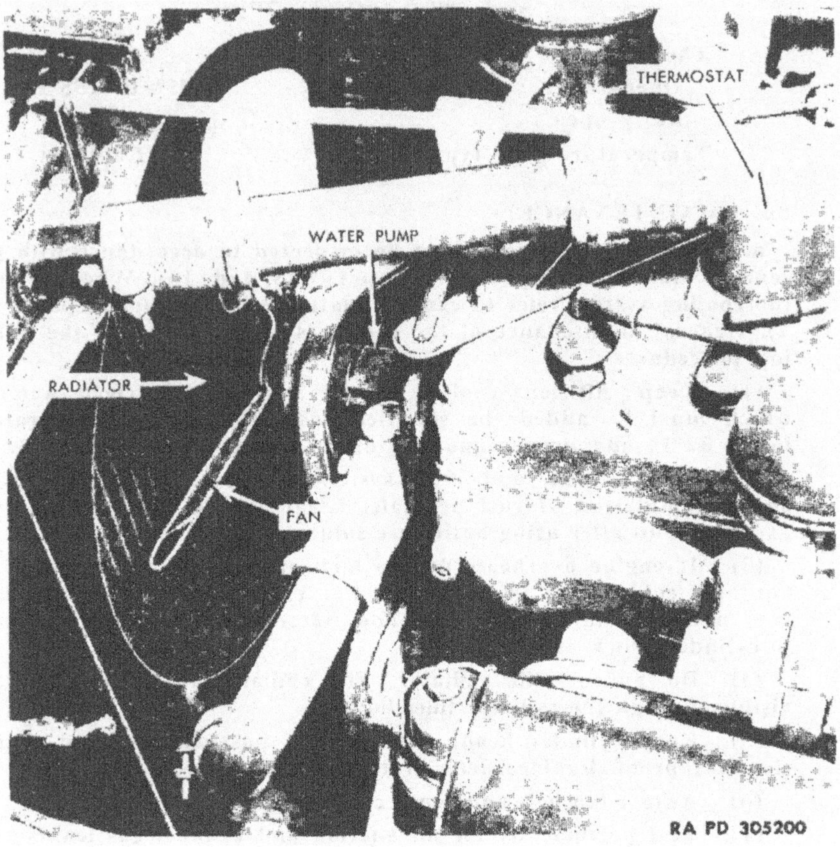

Figure 44—Cooling System

fill system to bottom edge of baffle in top of radiator. Replace pressure type cap with plain cap, and tighten securely (airtight). Attach length of rubber tubing to lower end of overflow pipe (this connection must be airtight). Run engine, with transmission in neutral, at a moderate speed until warmed up. Place tubing in glass container of water, and without changing engine speed, watch for bubbles in water. Continuous appearance of bubbles indicates that air is entering coolant. The cause will be one of the above, and must be corrected.

d. Exhaust Gas Leakage Test. The exhaust gas leakage test is used to determine if gas is entering the coolant, possibly due to leaky cylinder block, cylinder head, or gasket. NOTE: *Make this test with engine cold.* Remove fan belt. Open radiator drain cock until coolant is below cylinder head water outlet. NOTE: *Determine by loosening three screws holding outlet to head.* Remove water outlet, and fill cylinder head with coolant until level is up to top of head. With transmission in neutral, start engine, "gun" it several times, and watch for bubbles in water. Appearance of bubbles indicates leakage from one of the above conditions, which must be corrected. Replace leaky gasket; report other causes to higher authority.

e. Cleaning and Flushing Procedure. This procedure is used to clean out loose rust. Run engine at moderate speed to stir up loose rust. Drain cooling system. Close drain cocks, and fill system with specified cleaning compound. Install radiator cap. Operate engine as directed for prescribed solution. Stop engine and completely drain system by opening both drain cocks. To flush system close drain cocks, fill system with water, run engine until warmed up again, or run water through system, then completely drain. Close drain cocks, refill system, and add inhibitor corrosion compound to prevent formation of rust and scale. Inhibitor compound must be renewed periodically to be effective.

81. RADIATOR.

a. Description. The radiator assembly (fig. 44) consists of the fin-and-tube type core with a coolant tank at the top, and a sediment tank at the bottom. It is located in the conventional place at the front of the vehicle. A pressure-type filler cap maintains up to 4^-pound pressure to give better engine efficiency, and prevent evaporation of the coolant. When the engine is warm, release pressure by turning cap slightly before removal.

b. Removal. Remove the radiator filler cap. Open the drain cock on bottom of radiator outlet and right front corner of cylinder block to drain system. Remove radiator stay rod nut at front end. Loosen hose clamp on upper hose at front end. Loosen hose connection at water pump. Remove two radiator hold-down nuts and lift off radiator, then remove drain cock and pads.

c. Installation. Install drain cock and place pads on bracket; set radiator in place. NOTE: *A light coating of grease in hose connections will facilitate assembly.* Install stay rod and tighten lock nut. Tighten upper and lower hose clamps. Install hold-down nuts and bond strap. Close radiator drain cock and cylinder block drain cock. Fill radiator, install cap, and check system for leaks. Start engine, check coolant level after engine is warmed up, and fill, if needed, to proper level.

82. WATER PUMP.

a. Description. The water pump (fig. 44) is of the centrifugal-impeller, type and is located in the front end of the cylinder block. A double-row ball bearing is integral with the shaft, and is packed with lubricant when it is made. No lubrication attention is required. A packless self-sealing gland is used to prevent water leakage, and requires no attention. The water pump with the generator and fan is driven by a belt from the engine crankshaft fan pulley.

Figure 45—Fan Belt Deflection

b. **Removal.** Open the radiator drain cock in outlet pipe at lower left corner of radiator, and also drain cock in right side of cylinder block at forward end. Remove radiator filler cap. Pull up on handle of generator brace to loosen fan belt, and slip off belt. Loosen pump hose clamp and remove hose. Remove fan blade screws. Remove screws holding water pump in cylinder block, and remove water pump.

c. **Installation.** Check condition of pump to cylinder block gasket. Replace if damaged. Install pump in cylinder block, and tighten with screws. Install fan and screws. Install fan belt, and pull out generator until brace drops into position. Attach hose connection. Fill radiator, install cap, and check system for leaks. Start engine and check radiator coolant level after engine is warmed up.

83. FAN BELT.

a. **Description.** The fan belt (fig. 45) is of the V-type and drives the fan, water pump, and generator. Proper adjustment is necessary for efficient operation and maximum life of belt. Do not adjust the belt extremely tight, causing excessive wear on water pump and generator bearings.

b. Removal. Pull up on generator brace handle, and move generator toward engine as far as it will go. Remove belt from generator, pump, and crankshaft pulleys, and lift over fan blades.

c. Installation and Adjustment. Place belt over fan and crankshaft pulleys, then over the generator pulley. Pull generator out until

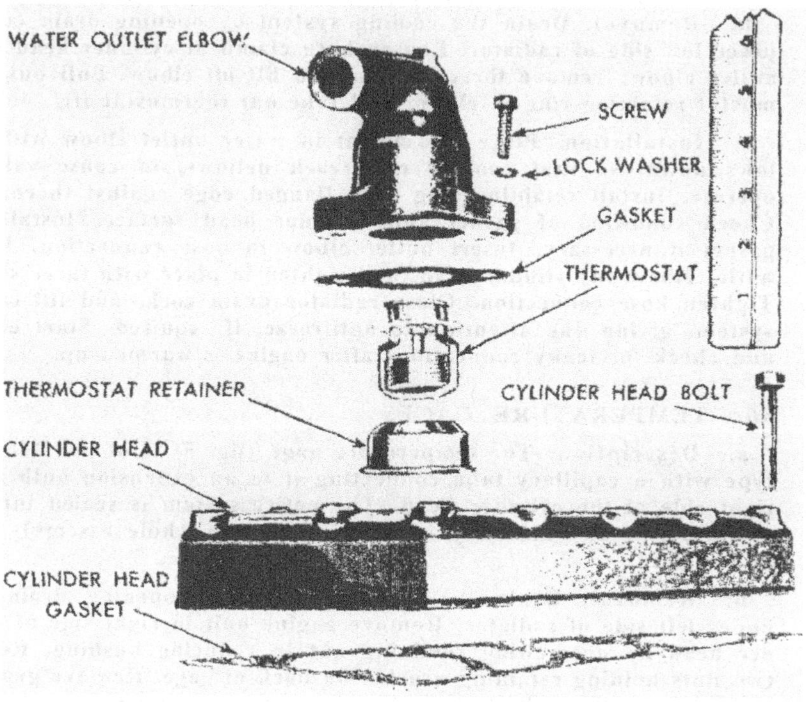

Figure 46—Thermostat, Disassembled

generator brace locks in position. To adjust fan belt tension, loosen generator brace nut, and move generator until fan belt has about one-inch deflection midway between fan and generator pulleys (fig. 45), then tighten nut.

84. FAN.

a. Description. A four-blade, 15-inch fan (fig. 44) draws air through the radiator core. It is mounted on the front end of the water pump shaft and is driven from the crankshaft by the same belt which drives the generator.

b. **Removal.** Remove four screws holding fan on fan pulley, and lift fan out of shroud.

c. **Installation.** Place fan in position on fan pulley, install four screws, and tighten securely.

85. THERMOSTAT.

a. **Description.** The thermostat (fig. 44) is of the bellows-type, and located in the water outlet elbow on top of the cylinder head. It is designed to open between 140°F and 155°F, and is fully open at 170°F.

b. **Removal.** Drain the cooling system by opening drain cock at lower left side of radiator. Loosen hose clamp at cylinder head water outlet elbow; remove three screws, and lift off elbow. Pull out thermostat retaining ring in elbow, and take out thermostat (fig. 46).

c. **Installation.** Place thermostat in water outlet elbow with bellows down, so that coolant can reach bellows, to cause valve to operate. Install retaining ring with flanged edge against thermostat. Check condition of gasket and cylinder head surface. Install new gasket if necessary. Insert outlet elbow in hose connection. Install outlet elbow on cylinder head, and tighten in place with three screws. Tighten hose connection. Close radiator drain cock, and fill cooling system, giving due attention to antifreeze, if required. Start engine, and check for leaky connections after engine is warmed up.

86. TEMPERATURE GAGE.

a. **Description.** The temperature gage (fig. 5) is of the Bourdon-type with a capillary tube connecting it to an expansion bulb in the right side of the cylinder head. The entire system is sealed into one assembly; if any difficulty is experienced, the whole assembly must be replaced.

b. **Removal.** Drain the cooling system by opening drain cock under left side of radiator. Remove engine unit in right side of cylinder head by unscrewing retaining nut in reducing bushing. Remove two nuts holding retaining bracket on back of gage. Remove grommet around tube through dash and draw out gage assembly, tube, and engine unit through hole in panel.

c. **Installation.** Insert engine unit through hole in panel, mounting bracket, and dash. Position gage correctly, install mounting bracket and nuts; tighten securely. Insert grommet in dash. Install engine unit, and tighten retaining nut. Fill cooling system, and check for leaks.

Section XVIII

STARTING SYSTEM

87. DESCRIPTION AND DATA.

a. Description. The starting system (fig. 47) is a 6-volt system. It consists of the starting switch, cranking motor, and the cables through which they are connected to the battery of the electrical system. When the starting switch is pressed with the right foot, the battery current energizes the cranking motor, causing the armature to turn. This results in the inertia-type cranking motor gear engaging the teeth on the outer circumference of the flywheel. When the engine starts, the gear is automatically thrown out of engagement with the flywheel teeth.

b. Data.

System voltage 6
Cranking motor
Normal engine cranking speed . . . . . 185 rpm
Make and model . . . . . Auto-Lite MZ-4113
Bearings 3
Brushes 4
Brush spring tension . . . . . 42 to 53 oz
Drive (direct) . . . . . R.H. outboard Bendix
Starting switch
Make and model . . . . . Auto-Lite SW-4015

88. MAINTENANCE.

a. The cranking motor requires lubrication only at the forward end. Check tightness of mounting screws. Wire terminals must be in good condition, clean, and tight. As part of the starting system, check the battery condition periodically. Clean the cranking motor drive periodically (par. 89).

89. CRANKING MOTOR.

a. Description. The cranking motor is a 6-volt, four-brush type, located at the right-rear side of the engine. The drive is transmitted to the engine from an inertia-type gear to the teeth on a removable ring gear on the circumference of the flywheel. Rotation of the cranking motor shaft causes the pinion of the cranking motor drive to advance and mesh with the gear on the flywheel. After the engine starts, and the speed of the flywheel exceeds that of the cranking motor, the flywheel disengages the pinion automatically. A removable cover at the front end of the cranking motor permits inspection of the brushes and commutator.

b. Removal. Remove cable from motor. Remove screw from front support bracket, and remove two attaching screws at flywheel. Remove cranking motor. Remove front bracket.

c. Installation. Clean cranking motor gear. *Do not oil drive.* Install front bracket and place cranking motor in position on engine. Insert attaching screws, also bracket screw, and tighten securely. Attach cable.

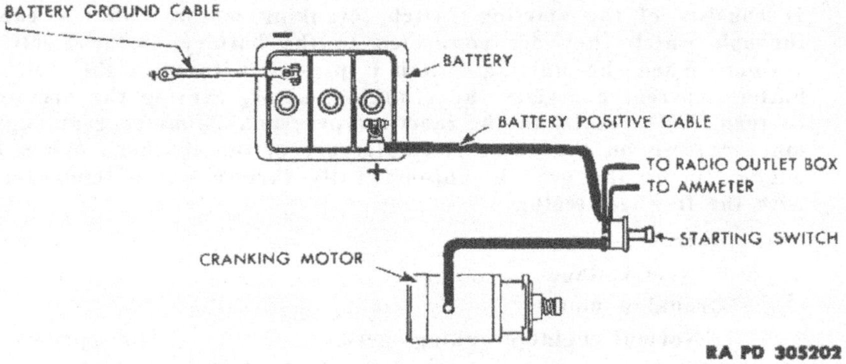

Figure 47—Starting System Circuit

90. STARTING SWITCH.

a. Description. The starting switch is a push-button type operated with the right foot, and located to the right of the accelerator treadle on the toeboard (fig. 5). Press the button to close the switch and cause the cranking motor to operate.

b. Removal. Remove battery cable at negative post of battery. Remove air cleaner (par. 73). Remove cable terminal nuts on switch, and remove cables and wires. Remove two screws holding switch to toeboard, and remove switch from underneath.

c. Installation. Remove cable terminal nuts, place switch in position under toeboard, and tighten in place with two screws. Attach cables and wires. NOTE: *Battery positive cable and wires to radio outlet box, filter, and ammeter are attached to top terminal.* Install air cleaner (par. 73), and connect battery cable to negative post on battery.

Section XIX

GENERATING SYSTEM

91. DESCRIPTION AND DATA.

a. Description. The generating system (fig. 49) is a 6-volt system, single-wire, ground-return type. This system consists of the generator, regulator, and wires connecting it to the ammeter. For information concerning the battery and lighting system refer to paragraph 95. The system develops current to keep the battery charged, and furnishes current for ignition, lighting, and other electrical accessories if the engine operation is sufficient. The regulator governs the generator output in accordance with the condition of the battery, and the requirements of the other electrical units used in the operation of the vehicle.

b. Data.

System voltage	6 to 8
Generator:	
Make and model	Auto-Lite GEG-5101D
Ground polarity	Negative
Controlled output	40 amps
Rotation (drive end)	Clockwise
Control	Current voltage regulator
Brushes	2
Output	8.0 amps; 7.6 volts; 955 rpm
	40.0 amps; 7.6 volts; 1460 rpm
	40.0 amps; 8.0 volts; 1465 rpm
Regulator:	
Make and model	Auto-Lite VRY-4203 A
Type	Current voltage
Volts	6
Amperes	40
Ground polarity	Negative

92. MAINTENANCE.

a. The generator requires attention to lubrication of those generators provided with oilers. Properly adjust the drive belt. Check mounting screws and bracket attached to engine. Generator brace must be in position. Wire terminals must be in good condition, clean, and tight. External connecting wires with damaged or cracked insulation must be replaced. Bond straps must be cleaned and securely tightened. If trouble is experienced with the generator, report to higher authority.

93. GENERATOR.

a. Description. The generator is located on the right side of the engine at the forward end (fig. 49). It is a 6-8 volt, shunt-wound, two-brush unit rotating clockwise as viewed from the drive pulley end. It has a controlled output of 40 amperes, and is air-cooled by a fan built into the drive pulley. Air is drawn in at the back of the generator, and discharged at the fan.

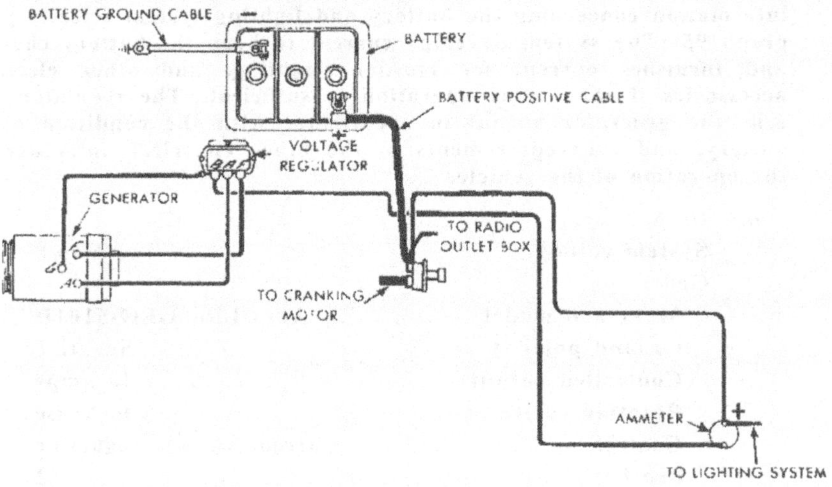

RA PD 305203

Figure 48—Generating System Circuit

b. Removal. Remove generator brace spring (fig. 10) and slip fan belt off pulley. Disconnect wires at generator; remove radio filter and ground strap. Remove two generator support bolts and remove generator.

c. Installation. Place generator in position, and install a drift or pin through front hole as a pilot. Install flat washer and bushing at rear support, and install bolt through hole. Remove drift, and install washer and bolt in front hole. Install generator brace spring. Install fan belt, and adjust if necessary (par. 83). Attach wires, filter, and ground strap. Start engine and check charging rate on ammeter in instrument panel (fig. 5).

94. REGULATOR.

a. Description. The regulator (fig. 49) automatically governs the output of the generator in accordance with the condition of the battery and the current requirements in the operation of the vehicle, thus, when the battery is low the output is increased, and as the

Figure 49—Generator and Regulator

battery becomes fully charged the generator develops less current to avoid overcharging. The regulator is mounted inside the splasher of the right front fender. It is a precision instrument, sealed at the factory, and no attempt should be made to adjust it. The regulator consists of three separate units: the cut-out or circuit breaker, the voltage regulator, and the current regulator. The circuit breaker automatically closes the circuit between the generator and the battery when the generator voltage rises above that of the battery, and opens the circuit when the generator current falls below that of the battery.

The voltage regulator governs the generator so that it will not develop more voltage than the value for which the voltage regulator is set. The current regulator controls the generator current (amperage) output so that it will develop enough to keep the battery charged (providing the engine is run sufficiently), and also prevent damage to the generator due to an overload.

b. Removal. Remove battery cable at battery negative post. Remove wires from regulator terminals. Mark wires to assure correct installation. Remove bond strap, and remove four bolts holding regulator to fender.

c. Installation. Place regulator in position on fender, and tighten securely with four bolts. Install bond strap. Connect wires to proper terminals. Install battery cable on post, and tighten securely. Start engine and check charging rate of ammeter in instrument panel.

Section XX

BATTERY AND LIGHTING SYSTEM

95. DESCRIPTION AND DATA.

a. Description. The lighting system (fig. 50) functions on 6 volts supplied by a three-cell storage battery. The system consists of two service headlights, two blackout headlights, one blackout driving light, service and blackout stop and taillights, two instrument panel lights, and operating switches and battery. The entire lighting system is controlled by the blackout (main) light switch on the instrument panel. When the blackout light switch is set in the proper operating position, other light switches such as blackout driving light, instrument panel, stop lights and dimmers, are controlled by the respective switches. The lighting system is protected by a thermal-type fuse on the back of the blackout switch.

b. Data.

Battery:	
Make and model	Auto-Lite-TS-2-15 Willard SW-2-119
Volts	6
Plates per cell	15
Capacity (ampere hours)	116
Length	10 in.
Width	7 in.
Height	8 5/16 in.
Ground terminal	Negative
Wiring system:	
Volts	6
Wiring identification reference	Fig. 50

RA PI

Figure 50—Wiring System—Phantom View

¼-TON 4 x 4 TRUCK (WILLYS-OVERLAND MODEL MB and FORD MODEL GPW)

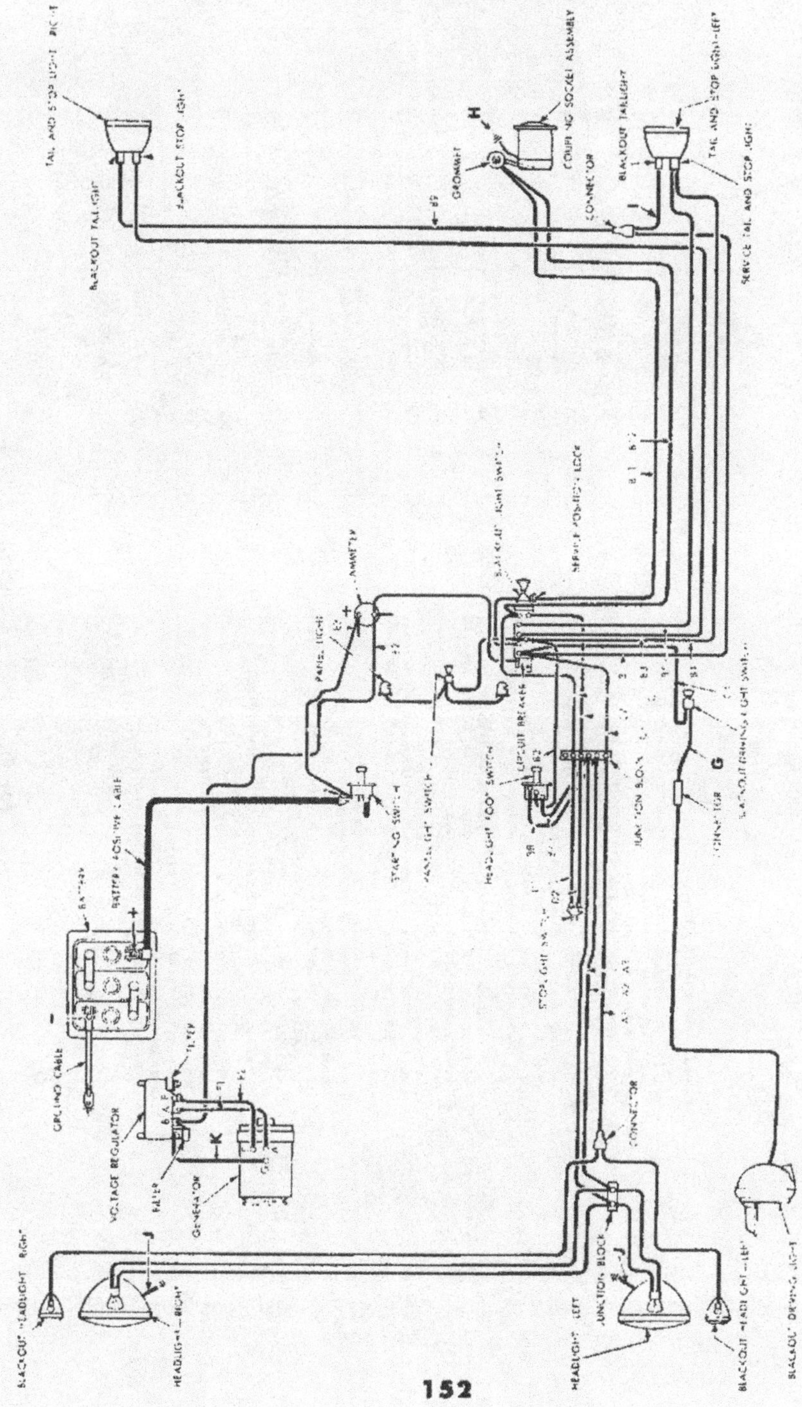

RA PD 305215

Figure 51—Lighting Circuits

	GAGE	COLOR
A—HEADLIGHT WIRING HARNESS		
A-1 BLACKOUT HEADLIGHT JUNCTION BLOCK TO JUNCTION BLOCK	14	YELLOW 2 BLACK TR.
A-2 HEADLIGHT JUNCTION BLOCK TO JUNCTION BLOCK (UPPER BEAM)	12	RED—3 WHITE TR.
A-3 HEADLIGHT JUNCTION BLOCK TO JUNCTION BLOCK (LOWER BEAM)	14	BLACK—2 WHITE TR.
B—BODY WIRING HARNESS—LONG		
B-1 LIGHT SWITCH TERMINAL "B H T" TO BLACKOUT TAIL CONNECTION	14	YELLOW—2 BLACK TR.
B-2 LIGHT SWITCH TERMINAL "H T" TO FOOT DIMMER SWITCH CENTER TERMINAL	12	BLUE 3 WHITE TR.
B-3 LIGHT SWITCH TERMINAL "B S" TO BLACKOUT STOPLIGHT	14	WHITE—2 BLACK TR.
B-4 LIGHT SWITCH TERMINAL "H T" TO SERVICE TAILLIGHT AND INSTRUMENT LIGHT SWITCH	14	BLUE—2 WHITE TR.
B-5 LIGHT SWITCH TERMINAL "S" TO SERVICE STOPLIGHT	14	RED—2 WHITE TR.
B-6 HORN CIRCUIT BREAKER TO HORN	14	BLACK 2 RED TR.
B-7 JUNCTION BLOCK TO FOOT DIMMER SWITCH (LOWER BEAM)	14	BLACK 2 WHITE TR.
B-8 JUNCTION BLOCK TO FOOT DIMMER SWITCH (UPPER BEAM)	12	RED—3 WHITE TR.
B-9 CONNECTOR TO BLACKOUT TAILLIGHT	14	YELLOW 2 BLACK TR.
B-11 LIGHT SWITCH TERMINAL "T T" TO COUPLING SOCKET TERMINAL "T L"	14	GREEN—2 BLACK TR.
B-12 LIGHT SWITCH TERMINAL "S S" TO COUPLING SOCKET TERMINAL "S L"	14	RED 2 BLACK TR.
C—BODY WIRING HARNESS—LEFT SIDE—SHORT		
C-1 JUNCTION BLOCK TO LIGHT SWITCH TERMINAL "S S"	14	RED—2 WHITE TR.
C-2 JUNCTION BLOCK TO LIGHT SWITCH TERMINAL "S W"	14	GREEN—2 BLACK TR.
C-3 JUNCTION BLOCK TO LIGHT SWITCH TERMINAL "B H T"	14	YELLOW—2 BLACK TR.
C-4 BLACKOUT LIGHT SWITCH TO LIGHT SWITCH TERMINAL "B H T"	14	BLACK 2 WHITE TR.
D—CHASSIS WIRING HARNESS—LEFT		
D-1 STOPLIGHT SWITCH TO JUNCTION BLOCK	14	RED—2 WHITE TR.
D-2 STOPLIGHT SWITCH TO JUNCTION BLOCK	14	GREEN—2 BLACK TR.
E—BODY WIRING HARNESS—RIGHT		
F 1 COIL TO IGNITION SWITCH TO GASOLINE GAGE CIRCUIT BREAKER	14	BLACK—2 WHITE TR.
F-2 VOLTAGE REGULATOR TERMINAL "B H T" TO AMMETER	12	RED—3 WHITE TR.
E 3 STARTING SWITCH TO AMMETER	12	BLACK—3 WHITE TR.
E 4 AMMETER TO HORN CIRCUIT BREAKER	14	BLACK—2 RED TR.
F—GENERATOR TO VOLTAGE REGULATOR AND FILTER HARNESS		
F 1 GENERATOR TO REGULATOR ARMATURE	12	RED—3 WHITE TR.
F 2 GENERATOR TO REGULATOR FIELD	14	GREEN—2 BLACK TR.
G—BLACKOUT DRIVING LIGHT CONNECTOR TO SWITCH	14	BLACK 2 WHITE TR.
H—COUPLING SOCKET TO GROUND	14	BLACK—2 WHITE TR.
I—CONNECTOR TO BLACKOUT TAILLIGHT	14	YELLOW 2 WHITE TR.
J—HEADLIGHT GROUND	12	BLACK 3 WHITE TR.
K—GENERATOR TO VOLTAGE REGULATOR—GROUND		FLEXIBLE BRAIDED STRAP

RA PD 305215B

Legend for Figure 51—Lighting Circuits

Lamps:

Headlights	Sealed unit
Blackout headlights	Mazda No. 1245
Blackout driving light	Sealed unit
Taillights and stop lights	Sealed unit
Instrument lights	Mazda No. 51

Trailer connections:

Make	Wagner
Socket model	No. 3604
Plug model	No. 3544

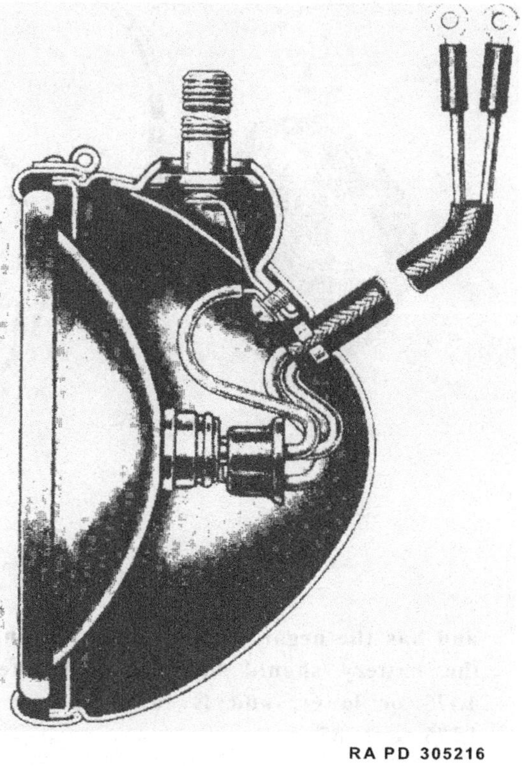

Figure 52—Headlight

RA PD 305216

96. MAINTENANCE.

a. The battery requires periodic checking of the proper level of electrolyte. Battery case must be kept clean. All vent plugs must be tight and breather holes kept open. Keep battery terminals and posts clean and securely tightened. Clean battery carrier when corroded. Tighten battery hold-down clamps. All light mounting screws and nuts must be kept clean and tight. Light lenses and reflectors must be kept clean and securely fastened. Headlights must be aimed properly. All loose and dirty electrical connections must be cleaned and tightened. All damaged or frayed wires must be replaced. Bond straps must be clean and securely tightened.

97. BATTERY.

a. **Description.** The battery (fig. 36) is a 6-volt, 116-ampere-hour storage battery consisting of three side-by-side cells of 15 plates each. The battery is located under the hood at the right side,

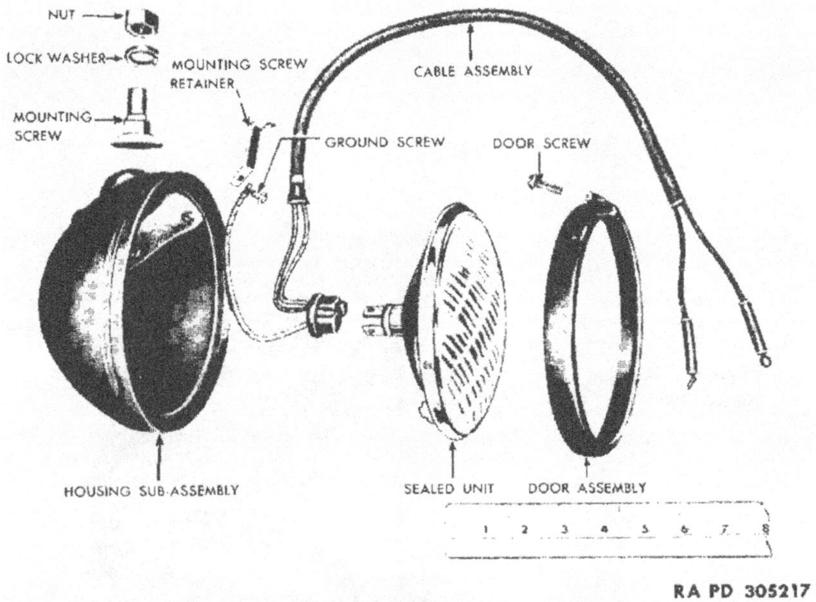

Figure 53—Headlight, Disassembled

and has the negative (small) post grounded. In normal temperatures the battery should be recharged when the specific gravity reads 1.175 or lower, and is fully charged when the specific gravity is 1.275 to 1.285.

b. **Removal.** Loosen bolts in cable terminals, and remove terminals from battery. Loosen two nuts holding frame on battery, and remove hold-down frame. Lift out battery.

c. **Installation.** Place battery in position with positive post to rear. Place hold-down frame in position, and tighten securely with nuts. Clean cable terminals; if necessary; grease and install on battery posts. Check level and specific gravity of electrolyte in battery. Electrolyte level must be 1/2 inch above plates, and specific gravity must be 1.275 to 1.285 at 80°F.

98. WIRING SYSTEM.

a. Description. The entire vehicle wiring system is shown in figure 50, and the circuits of the lighting system are shown in figure 51. A single-wire system is used, and the negative terminal of the battery is grounded. A single cable connects the battery positive post to the upper terminal of the starting switch, to which are also attached wires running to the radio outlet box, filter, and right-hand terminal of the ammeter, where the horn wire is connected. From the left-hand ammeter terminal wires run to the voltage regulator, ignition switch, and blackout (main) light switch, which controls

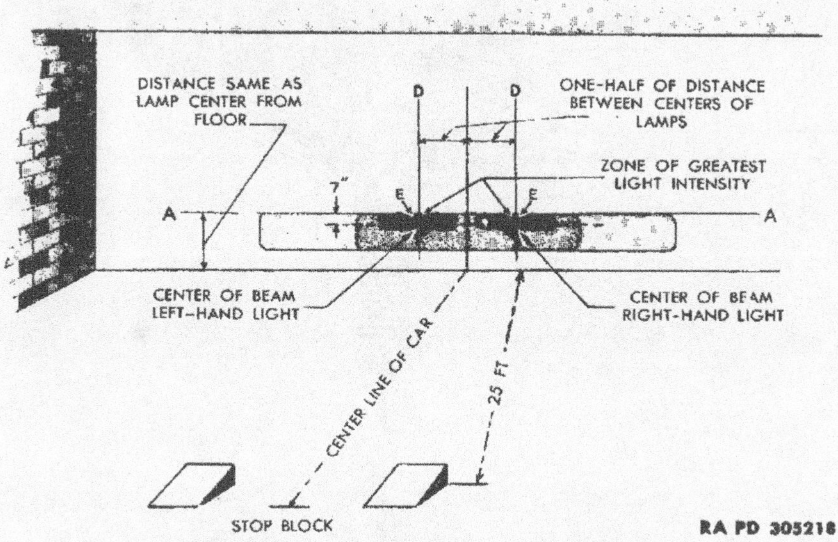

Figure 54—Headlight Aiming Chart

all lights in the lighting system. The wiring system consists of the harnesses and wires which are identified by colors and tracers as shown in figure 50, and described in the accompanying legend.

b. Removal. When necessary to renew wires in the wiring system, remove grounded (negative) cable at battery. Note location of wires; disconnect wires at terminals; loosen clips along harness, and remove wires.

c. Installation. When installing wires and harnesses see that all connections are correctly located and secure. Double-check location and tightness of connections; install retaining clips along wires.

99. HEADLIGHTS.

a. **Description.** Two headlights (fig. 52) are mounted at the front of the vehicle and are protected by the radiator guard. Each headlight is mounted on a hinged bracket. By loosening the wing nut, the light can be swung over and used as a trouble light in the engine compartment. The headlights are of the double filament sealed beam-unit type consisting of reflector, lamp, and lens, which can be

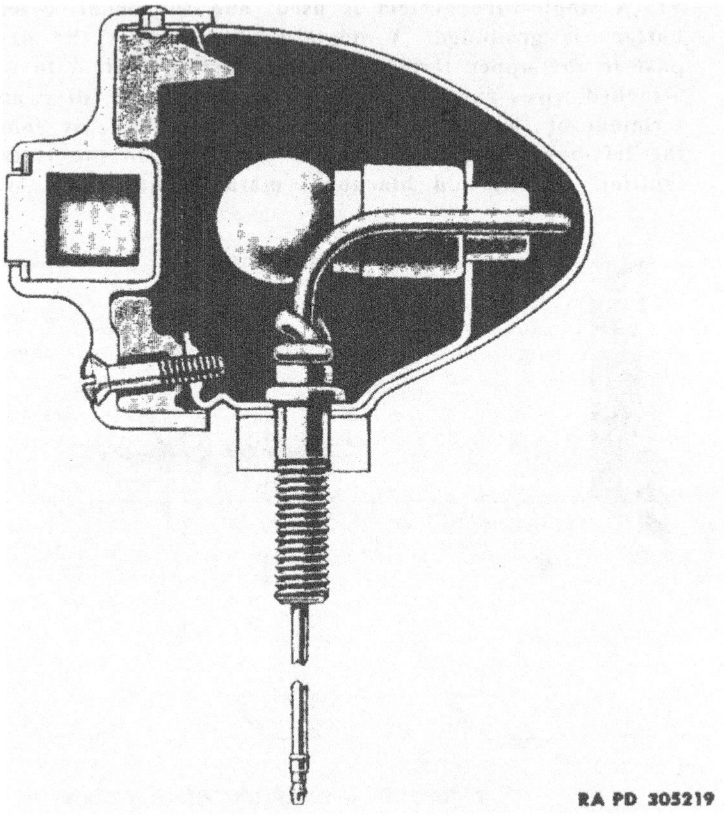

Figure 55—Blackout Headlight

replaced only as a unit (Mazda No. 2400). The headlights light when the blackout lighting switch is in service position. The upper and lower beams are controlled by the foot-operated dimmer switch. The lower beam filament is positioned slightly to one side of the focal point deflecting the beam slightly to illuminate the right side of the road.

b. **Removal of Sealed Beam-unit.** Remove door clamp screw and remove door (fig. 53). Remove sealed beam assembly, and pull connector from back of unit.

c. **Installation of Sealed Beam-unit.** Attach connector to back of unit. Place unit in position in headlight housing. Install door, and secure in place with screw.

d. **Removal of Headlight.** Disconnect headlight wires at junction block on left fender splash under hood. Remove clip on radiator grille (two on right headlight wire). Remove nut on headlight. Loosen headlight support wing nut, raise support, and remove headlight.

e. **Installation of Headlight.** Mount the headlight on the support. Install wire clip on grille (two on right headlight wire). Connect wires to junction block (red wire on bottom terminal; black wire on top terminal). Lower support, and fasten with wing nut. Aim headlight (subpar. f below).

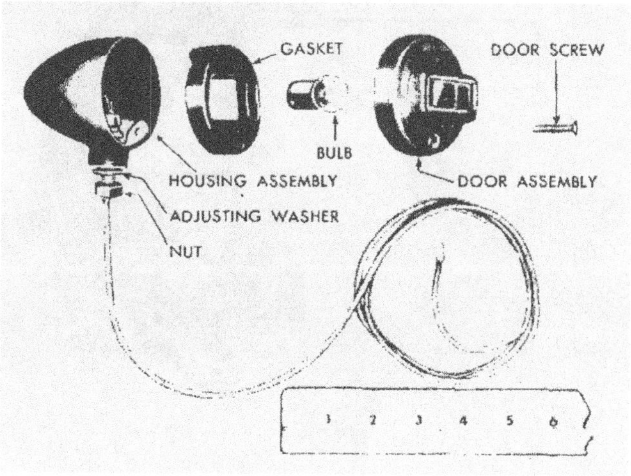

Figure 56—Blackout Headlight, Disassembled

RA PD 305220

f. **Aiming Headlights.** Aim headlights by using headlight aimer, aiming screen, or wall shown in figure 54. NOTE: *Use a light background with black center line for centering vehicle on screen. Mark two vertical lines on each side of center line equal to the distance between headlight centers. Mark screen 7 inches less than the height of centers of headlights.* Inflate all tires to recommended pressure (par. 3). Set vehicle so headlights are 25 feet away from screen and center line of vehicle is in line with center line of screen. To determine center line of vehicle, stand at rear and sight through windshield down across cowl and hood. Turn on headlight upper beam, cover one headlight, and observe location of upper beam on screen. Adjust headlight so that center of bright light area is on intersection of vertical and horizontal lines. Tighten headlight mounting nut. Cover headlight just aimed, and adjust other in same manner.

100. BLACKOUT HEADLIGHTS.

a. **Description.** There are two blackout headlights (fig. 55) with lens which permit only horizontal rays to pass through. These headlights are illuminated only when the blackout (main) light switch is in blackout position. Mazda No. 1245 lamps are used.

b. **Removal of Lamp.** Remove door screw in lower side of rim and remove door by pulling out on lower side (fig. 56). NOTE: *The door and lens are in one unit.* Push in on lamp, turn lamp to left and remove.

c. **Installation of Lamp.** Insert lamp in socket, push in and turn lamp to right. Replace door gasket, if damaged. Replace door and tighten door screw.

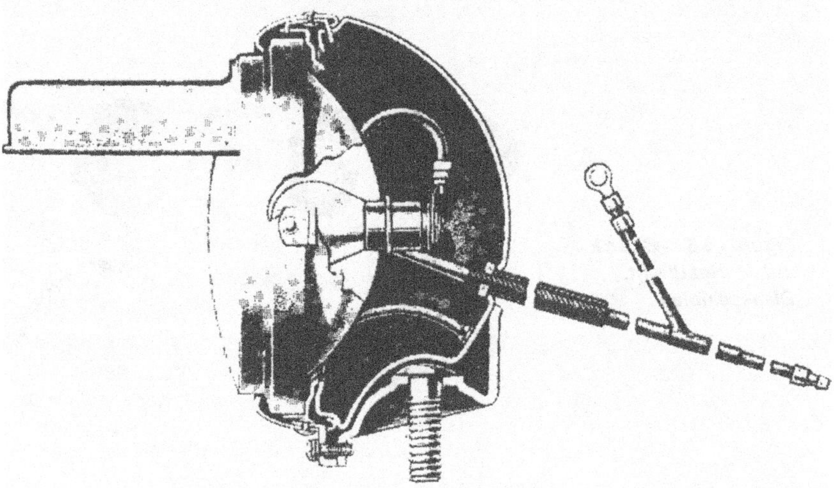

RA PD 305221

figure 57—Blackout Driving Light

d. **Removal of Blackout Headlight.** Pull wire out of connection just behind left headlight. (For right headlight remove three clips across bottom of grille.) Remove mounting nut by reaching in from the rear and lift out headlight.

e. **Installation of Blackout Headlight.** Put light in position and tighten in place with mounting nut. (Attach three clips for right headlight.) Attach wire in connector.

101. BLACKOUT DRIVING LIGHT.

a. **Description.** The blackout driving light (fig. 57) is a hooded light having a sealed unit which throws a horizontal diffused beam to illuminate any vertical object. This light is mounted on the left front fender, and is controlled by a push-pull type switch (marked "B.O. DRIVE") in the instrument panel. The blackout driving light can be illuminated only when the blackout (main) light switch is in blackout position.

b. Removal of Sealed Unit. Remove screw at bottom of light door (fig. 58). Pull bottom of door forward and up to remove. Disconnect socket wire, and remove door by releasing mounting ring.

c. Installation of Sealed Unit. Place sealed unit in door, and install unit mounting ring. Attach wire connection. Install door on housing, and tighten retaining screw.

d. Removal of Blackout Driving Light. Pull wire out of connector at dash. Remove three wire clips, and pull wire through fender. Remove mounting nut on bottom of light, and lift off light.

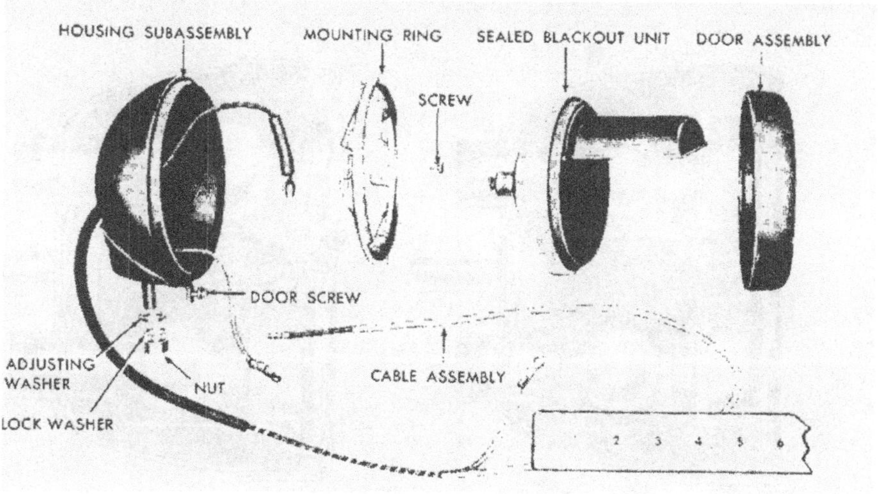

Figure 58—Blackout Driving Light, Disassembled

e. Installation of Blackout Driving Light. Install light on bracket and tighten nut; run wire through fender and splasher; attach three clips, and push wire into connector at dash.

f. Adjustment. The light is aimed with vehicle on a level surface and loaded. Hold a 4-foot wood stick on floor close to the light, and mark stick where bright spot appears. Move stick 10 feet ahead of light; bright spot should be 2.1 inches lower than mark.

102. TAILLIGHTS AND STOP LIGHTS.

a. Description. Two combination taillights and stop lights (fig. 59) are located in the rear panel of the body. Each light consists of two separate units in a housing. The left-hand light contains a combination service tail and stop light unit in the upper part, and a blackout taillight in the lower part. The upper unit consists of the taillight lens, gasket, reflector, and a 21-3-candlepower lamp. The lower unit consists of blackout lens, gasket, reflector, and 3-candlepower lamp. The right-hand light contains a blackout stop light unit in the upper part, and a blackout taillight in the lowei part. The upper unit consists of the blackout stop light lens, gasket, reflector, and 3-candlepower lamp. The lower unit is the same as the lower unit in the left light. When a lamp burns out, the unit must be replaced. These lights are controlled by the blackout (main) light switch.

b. Removal of Light Unit. Remove two screws in light door, and remove door (fig. 60). Pull each unit straight out of socket.

c. Installation of Light Unit. Be sure that unit is correct type, and push into socket. Install light door and screws.

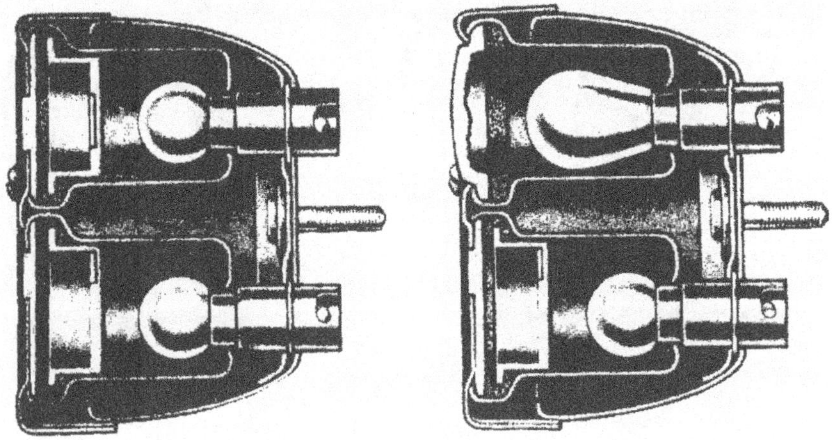

RA PD 305223

Fijure 59—Taillights and Stop Lights

d. Removal of Light. Reach up under body and disconnect wire connector; push in on connector, turn counterclockwise and pull connector out of socket. Remove two nuts holding light to bracket, and remove light.

e. Installation of Light. Place light in position and secure with nuts. Attach connectors. Double contact connector goes in upper socket of left light. Turn on blackout (main) light switch to blackout position to see if right blackout taillight (lower unit) lights; if not, interchange connectors in sockets.

103. INSTRUMENT PANEL LIGHTS.

a. Description. There are two instrument panel lights (fig. 5) located above instruments and on the outside of the instrument panel. They are controlled by the panel light switch when the blackout (main) light switch is in service position.

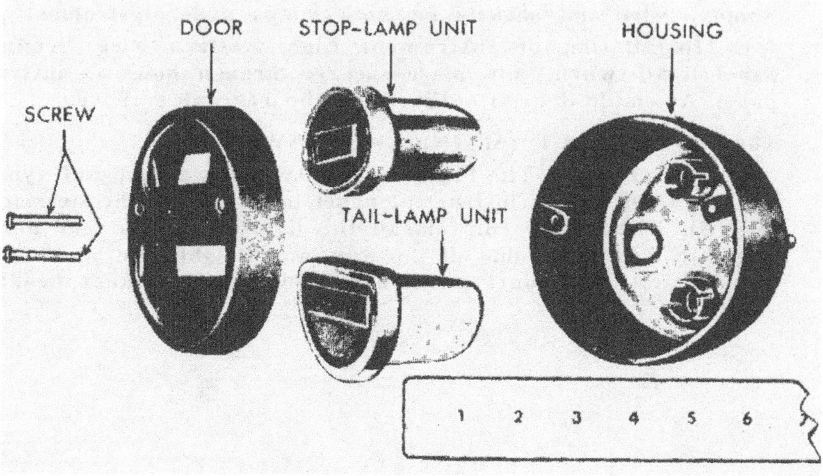

RA PD 305224

Figure 60—Taillights and Stop Lights, Disassembled

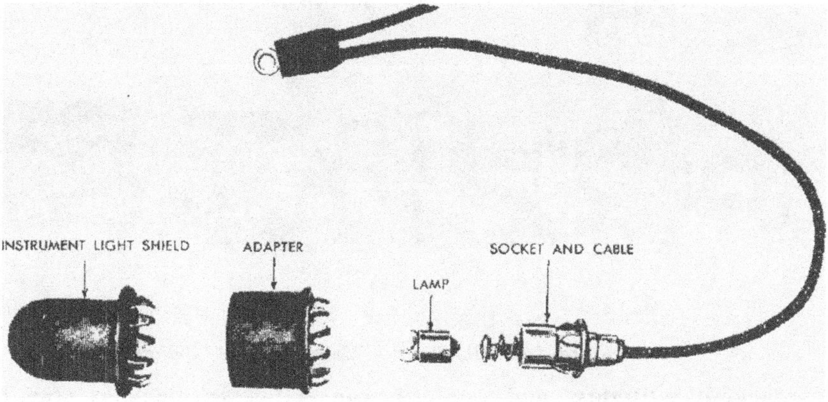

RA PD 305225

Figure 61—Instrument Light, Disassembled

b. Removal of Lamp. Pry off shield by using a sharp tool behind flange (fig. 61). Pull light socket out of shield; press in on lamp, turn lamp counterclockwise, and remove.

c. Installation of Lamp. Put lamp in socket, push lamp in, and turn clockwise. Push socket into shield, and push shield into instrument lamp adapter.

d. Removal of Instrument Light. Disconnect wire on back of light switch, and dismantle as outlined in subparagraph b above. Remove wire and sockets for both lamps under instrument panel.

e. Installation of Instrument Light. Attach wire terminal to panel light switch, and place sockets through holes in instrument panel. Assemble light as outlined in subparagraph c above.

104. BLACKOUT (MAIN) LIGHT SWITCH.

a. Description. The blackout light switch is a push-pull type (fig. 62), mounted in the instrument panel to the left of the steering gear (fig. 5). The switch controls all the lights, and has four positions (fig. 9). With the knob all the way in, all lights are off. Pull knob out to first (blackout) position to illuminate blackout headlights,

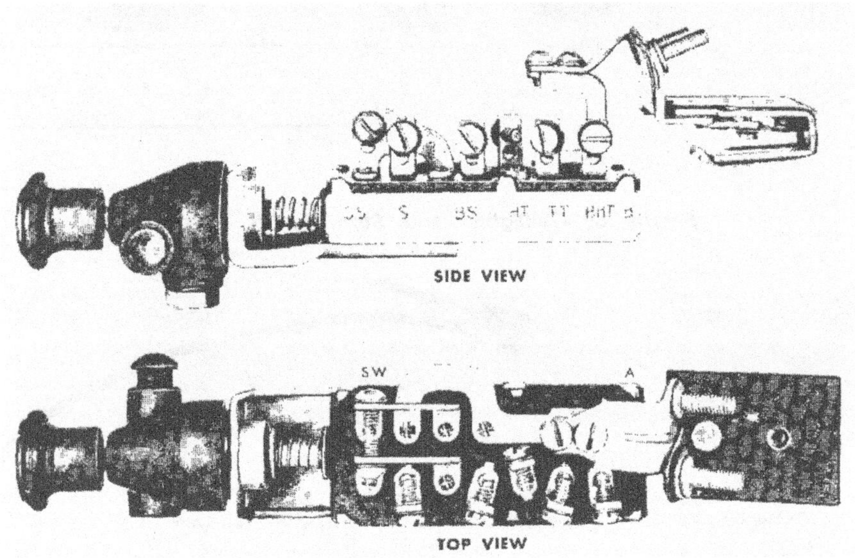

Figure 62—Blackout (Main) Light Switch

blackout taillights, and establish connection so blackout stop light will function when foot brake is applied. Press lockout control button, and pull switch knob out to second (service) position to illuminate service headlights and taillight, and to establish connections so service stop light will function when foot brake is applied. Pull knob out to third (service stop light) position, when vehicle is operated during daylight, to cause stop light only to function when foot brake is applied. A thermal-type fuse is mounted on the back of the switch having a bimetal spring, which causes a set of contact points to open and close, if a short circuit occurs in the lighting system.

b. Removal. Disconnect ground cable at negative post of battery. Loosen set screw in switch knob and unscrew knob. Loosen hexagon head screw at side of switch bushing on front of panel; press lockout control button, and pull off bushing. Remove mounting nut, and take switch out from under panel. Remove wire terminal screws. As wires are removed, mark them for identification.

c. Installation. Attach wires to proper terminals. NOTE: Swirch *terminals are marked for easy identification.* Wires are to be attached as outlined below:

Figure 63—Light Switch

RA PD 305227

Terminal	Circuit	Wire Color
A	... Extra	Not used
B	... Ammeter	Red-white tr.
BHT	... Blackout headlights	Yellow-black tr.
	Blackout taillights	Yellow-black tr.
	Blackout driving light	Black-white tr.
BS	... Blackout stop light	White-black tr.
HT	... Service headlight	Blue-white tr.
	... Service taillight	Blue-white tr.
	... Panel lights	Blue-white tr.
S	... Service stop light	Red-white tr.
SS	...Blackout driving light	Red-white tr.
	... Trailer coupling socket	Red-black tr.
SW	... Stop light	Green-black tr.
TT	... Trailer coupling socket	Green-black tr.

Install switch in panel and secure in place with mounting nut. Install bushings and tighten screw. Install knob and tighten set screw. Attach ground cable to battery.

105. PANEL AND BLACKOUT DRIVING LIGHT SWITCHES,

a. Description. The panel and blackout driving light switches

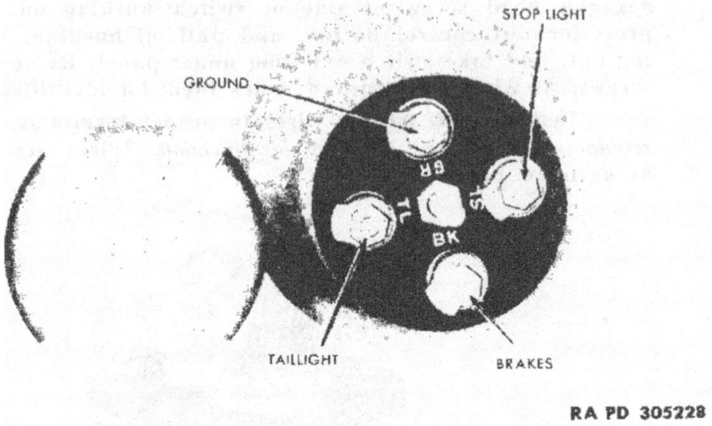

Figure 64—Trailer Socket Terminals

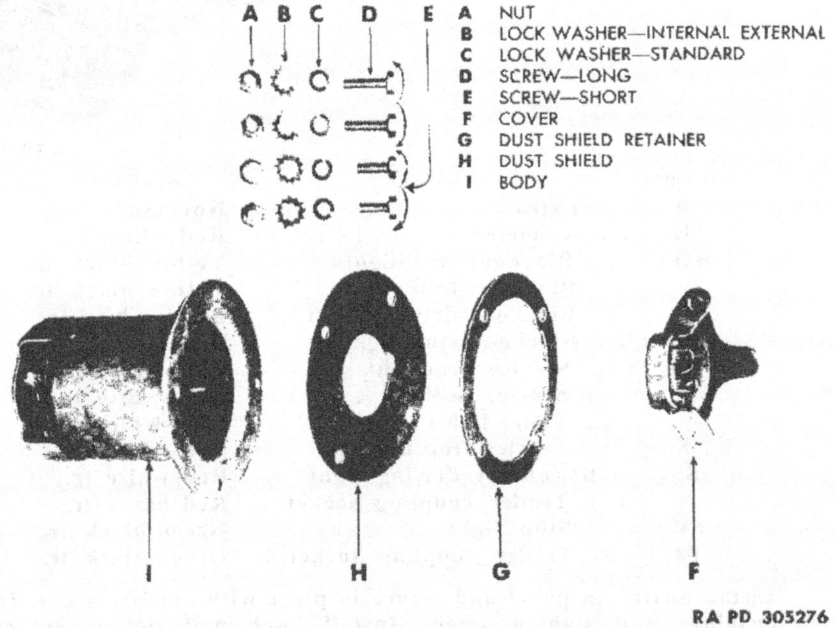

Figure 65—Trailer Socket, Disassembled

are push-pull (off-and-on) type (fig. 63). The panel light switch knob is marked "PANEL LIGHTS"; the blackout driving light switch is marked "B.O. DRIVE." The panel light switch controls these lights only when the blackout (main) light switch is in the service position. The blackout driving light switch controls that light only when the blackout (main) light switch is in blackout position.

b. Removal. As a precaution remove ground cable from battery. Loosen set screw in knob and unscrew knob. Remove mounting nut, and remove switch from in back of panel. Remove screws in wire terminals, and disconnect wires.

c. Installation. Attach wires to switch, install in place in panel, and secure with mounting nut. Install knob • and secure with set screw so that marking on knob is in correct position.

106. TRAILER CONNECTION.

a. Description. The trailer connection is a socket located in the body rear panel at the left side (fig. 2). The blackout (main) light switch controls the current to socket and trailer lights.

b. Removal. Remove equipment from left rear tool compartment. Remove screw at top edge of protecting cover, and remove cover. Remove four bolts holding socket in body panel. Pull out socket, and remove cover over the terminals. Remove wires, noting their proper place.

c. Installation. Attach wires to terminals. NOTE: *Attach green wire to terminal "TL"; attach red wire to terminal "SL"; attach small terminal of black wire to terminal "GR."* Install cover over terminals. To install socket in body, install the two long mounting screws in socket cover hinge. Place dust shield retainer ring over screws, followed by the dust shield with slot in center opening, opposite the hinge. Place cover against outside of body panel with the two screws through upper mounting holes so that cover opens upward. Place internal-external lock washers on each bolt. Install socket with drain hole down, and install lock washers and nuts loosely. NOTE: *Toothed lock washers must be installed as directed for a good ground connection.* Install the lower bolts with the lock washers in correct position. Install ground wire on a lower bolt, and tighten all four nuts holding socket in body panel. Install protecting cover, and secure in place with screw at top edge to complete the installation.

Section XXI

CLUTCH

Paragraph

107. DESCRIPTION AND DATA.

a. Description. The clutch (fig. 66) located between the engine and the transmission is a single plate, dry-disk type. The clutch consists principally of two units, the clutch driven plate which has a spring-center vibration neutralizes and the clutch pressure plate unit, which is bolted to the flywheel. The controlled pressure of the driven plate against the flywheel provides a means of engaging and disengaging the engine power to the transmission. A ball-type release bearing operates three clutch release levers to control the clutch. The release bearing is controlled by a rod and cable to the clutch pedal. This type clutch has only one service adjustment. This adjustment is for the foot pedal, and regulates the amount of free pedal travel. As the clutch facings wear, adjust the free pedal travel to three-quarters of an inch (fig. 68).

b. Data.

Type	Dry single plate
Torque capacity driven plate	132 foot-pounds
Make	Borg and Beck No. 11123
Size	7⅞ in.
Facings	1 woven and 1 molded
Diameter	5⅛ in. inside, 7⅞ in. outside
Thickness	⅛ in. (0.125)
Pressure plate:	
Make	Atwood No. TP-2B-7-1
Number of springs	3
Clutch release bearing type	Prelubricated ball
Clutch shaft bushing (in flywheel) size	I.D. 0.628 in.
Clutch pedal adjustment (free play)	¾ in.

108. MAINTENANCE.

a. The clutch requires attention to pedal adjustment. Pedal adjustment must be periodically checked, due to the natural wear of lining. Report grabbing or slipping condition of clutch to higher authority.

109. PEDAL ADJUSTMENT.

a. Adjust Pedal. Loosen clutch control cable adjusting yoke lock nut (figs. 67 and 68). Using a wrench, unscrew cable until clutch pedal has 3/4 inch free play. Tighten lock nut.

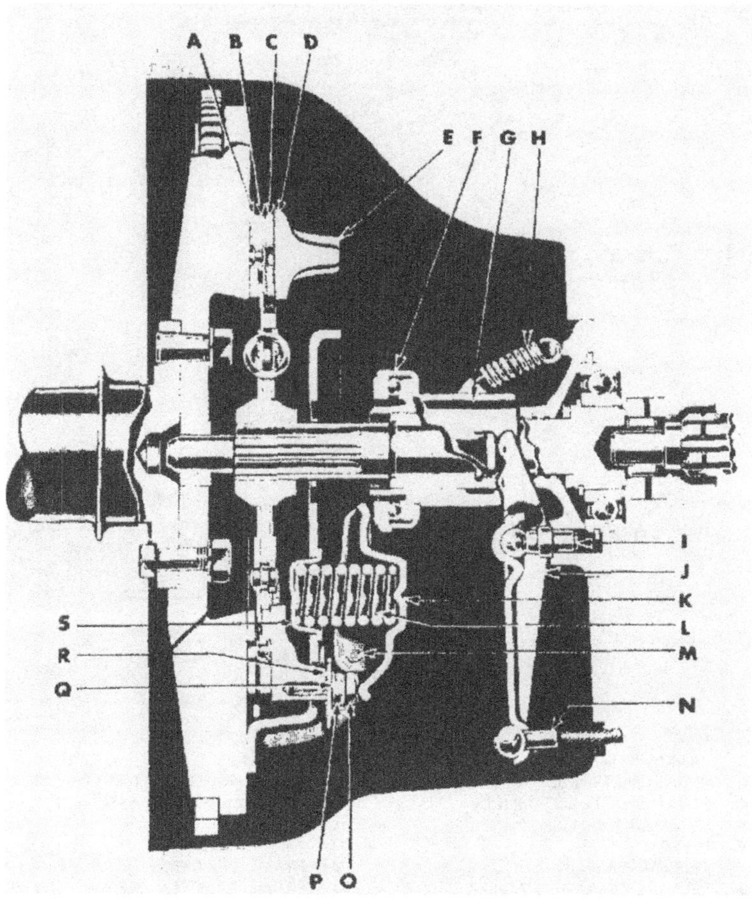

A FACING—FRONT
B DRIVEN PLATE ASSEMBLY
C FACING—REAR
D PRESSURE PLATE
E CLUTCH PRESSURE PLATE ASSEMBLY
F RELEASE BEARING
G RELEASE BEARING CARRIER
H RELEASE BEARING CARRIER SPRING
I CLUTCH CONTROL LEVER FULCRUM
J CLUTCH CONTROL LEVER
K CLUTCH LEVER
L PRESSURE SPRING
M CLUTCH LEVER PIVOT PIN
N CONTROL LEVER CABLE
O ADJUSTING SCREW
P ADJUSTING SCREW LOCK NUT
Q ADJUSTING SCREW WASHER
PRESSURE PLATE RETURN SPRING
S PRESSURE SPRING CUP

RA PD 305274

Figure 66—Clutch—Sectional View

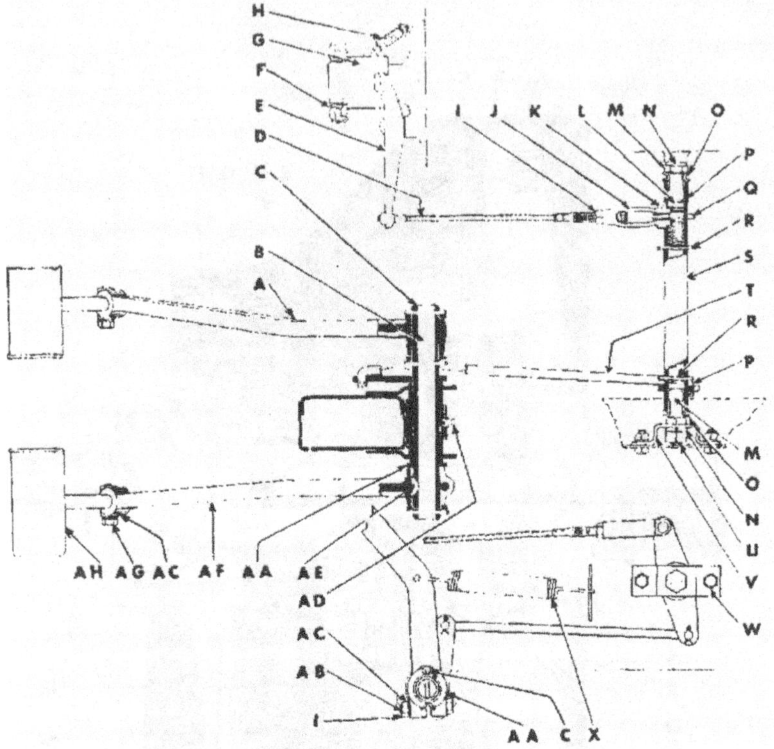

A	BRAKE PEDAL	Q	SPRING
B	PEDAL SHAFT ASSEMBLY	R	TUBE SPRING COTTER PIN
C	PEDAL SHAFT COTTER PIN	S	LEVER AND TUBE ASSEMBLY
D	LEVER CABLE	T	PEDAL ROD
E	LEVER	U	BALL STUD NUT
F	RELEASE BEARING	V	FRAME BRACKET
G	RELEASE BEARING CARRIER	W	FRAME BRACKET SCREW
H	RELEASE BEARING CARRIER SPRING	X	PEDAL RETRACTING SPRING
I	LEVER CABLE YOKE END LOCK NUT	AA	PEDAL SHAFT WASHER
J	LEVER CABLE YOKE END	AB	PEDAL CLAMP BOLT
K	LEVER CABLE CLEVIS PIN	AC	PEDAL CLAMP BOLT LOCK WASHER
L	TUBE WASHER	AD	PEDAL SHAFT HYDRAULIC GREASE FITTING
M	BALL STUD	AE	PEDAL TO SHAFT KEY
N	BALL STUD LOCK WASHER	AF	PEDAL
O	TUBE DUST WASHER	AG	PEDAL CLAMP BOLT
P	TUBE FELT WASHER	AH	PEDAL PAD

RA PD 305286

Figure 67—Clutch Control

110. REMOVAL.

a. Remove engine assembly (par. 60) or transmission and transfer case assembly (par. 115). NOTE: ***The easiest method is to remove the engine assembly.***

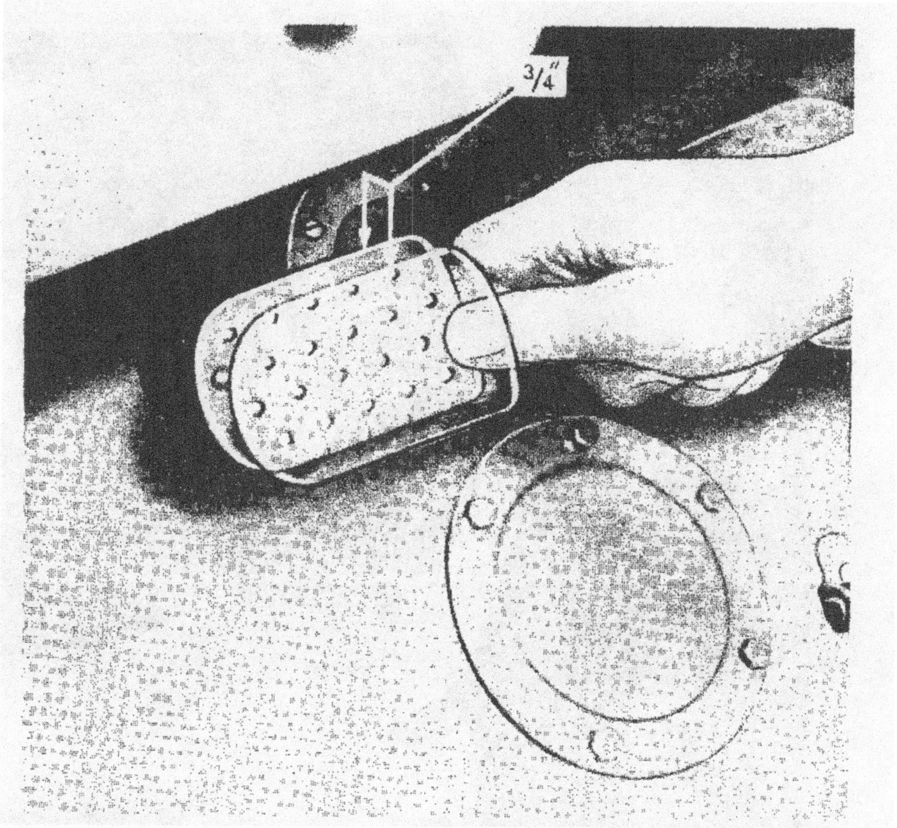

RA PD 305239

figure 68—Clutch Pedal free Travel

b. Mark clutch pressure plate and flywheel to assure correct position when installing. Loosen evenly and remove screws holding pressure plate to flywheel Remove pressure plate; remove driven plate.

111. INSTALLATION.

a. Clean flywheel and clutch. Install small amount of light grease in clutch shaft flywheel bushing. Install driven plate against flywheel with short end of hub toward flywheel. Install clutch pressure plate loosely with screws. Use a clutch shaft or clutch pilot arbor to aline driven plate, and tighten clutch pressure plate screws evenly. Remove pilot and check adjustment of clutch fingers (fig.

69), which should be 27/32 inch. To adjust clutch fingers, loosen lock nut on adjusting screws, and turn screws until measurements from face of fingers (release bearing contacts) to face of clutch bracket measures 27/32 inch; set lock nuts. Install engine (par. 61), or transmission and transfer case assembly (par. 116), as required.

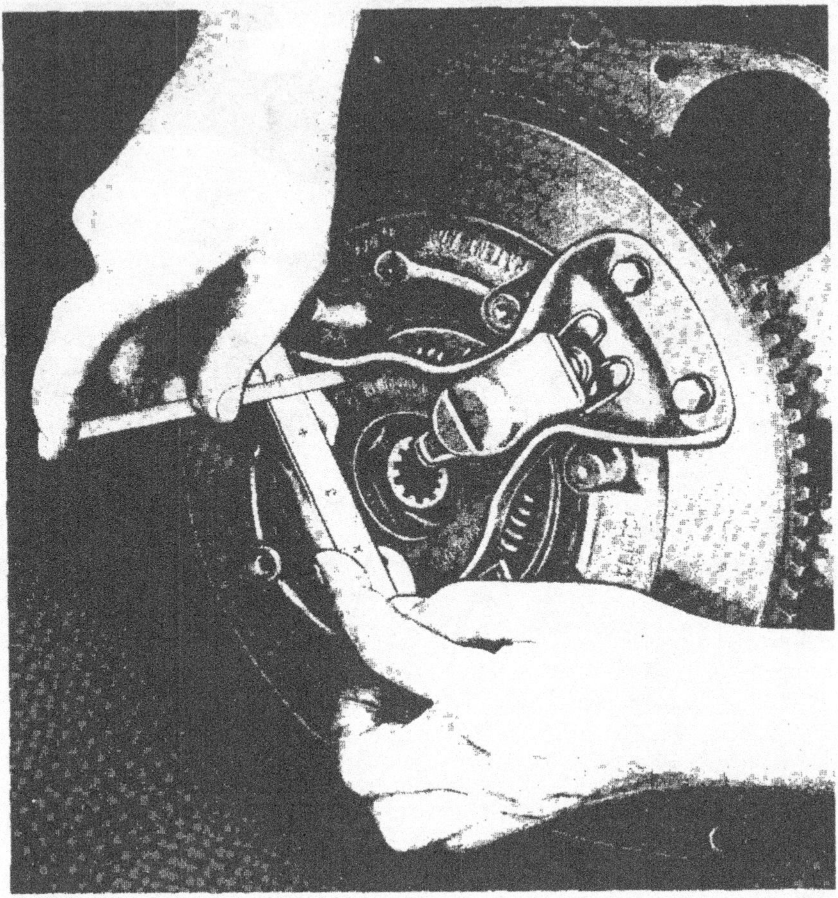

RA PD 305229

Figure 69—Clutch Finger Adjustment

112. CLUTCH RELEASE BEARING.

a. Removal. Follow procedure outlined in paragraph 60 for removing engine. After release bearing can be reached, unhook release bearing carrier spring, and pull off bearing and carrier (fig. 66). Press carrier out of bearing.

b. Installation. Press carrier into bearing. Slip bearing and carrier onto transmission bearing retainer, install release lever on fulcrum, and hook spring to bearing carrier. Follow the procedure outlined in paragraph 61 for completion of assembly.

Section XXII

TRANSMISSION

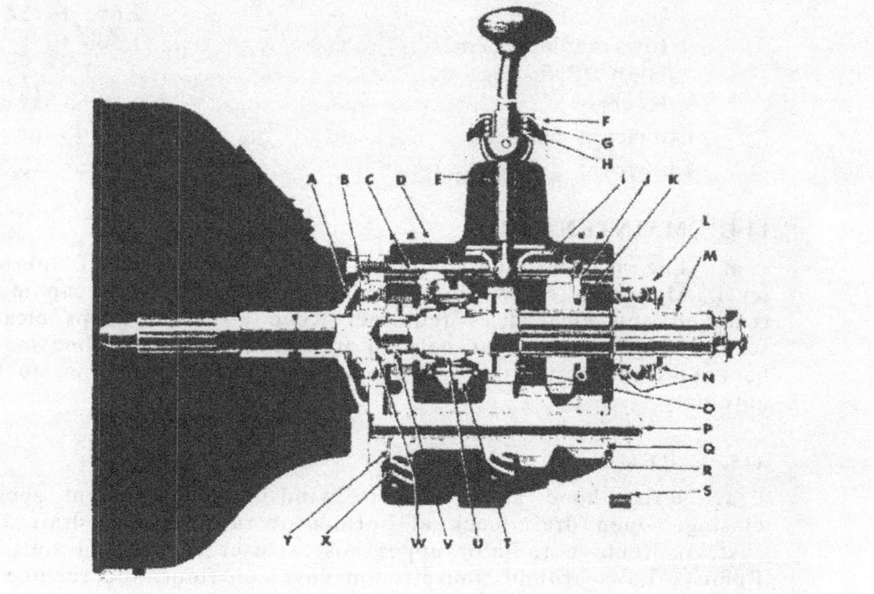

A MAIN DRIVE GEAR BEARING RETAINER
B MAIN DRIVE GEAR BEARING
C SHIFT RAIL—LOW AND REVERSE
D CONTROL HOUSING ASSEMBLY
E CONTROL LEVER ASSEMBLY
F CONTROL HOUSING CAP
G CONTROL HOUSING CAP WASHER
H CONTROL LEVER SUPPORT SPRING
I SHIFT PLATE
J LOW AND REVERSE SHIFT FORK
K LOW AND REVERSE SLIDING GEAR
I MAIN SHAFT BEARING
M MAIN SHAFT
N OIL RETAINING WASHER
O MAIN SHAFT SECOND SPEED GEAR ASSEMBLY
P COUNTERSHAFT
Q REAR COUNTERSHAFT THRUST WASHER (STEEL)
R REAR COUNTERSHAFT THRUST WASHER (BRONZE)
5 CASE
T SECOND AND DIRECT SPEED CLUTCH SLEEVE
U HIGH AND INTERMEDIATE CLUTCH HUB
V MAIN SHAFT PILOT ROLLER BEARING
W MAIN DRIVE GEAR
X COUNTERSHAFT GEAR ASSEMBLY
Y COUNTERSHAFT fHRUST WASHER-FRONT

RA PD 305285

Figure 70—Transmission—Sectional View

113. DESCRIPTION AND DATA.

a. **Description.** The transmission (fig. 70) is a selective, 3-speed, synchromesh type with synchronized second and high speed gears.

It is located in the power plant unit with the engine, clutch, and transfer case. Third speed is a direct drive through the transmission; all other speeds are through gears of various sizes to obtain the necessary gear reduction. The gears are shifted by a lever extending out of the top of the transmission, and through the floor at the right of the driver. For shifting instructions, refer to paragraph 5 **e**.

b. Data.

Make and model	Warner T-84-J
Type	Synchromesh
Speeds	3 forward—1 reverse
Ratios:	
Low (1st)	2.665 to 1
Intermediate (2nd)	1.564 to 1
High (3rd)	1 to 1
Reverse	3.554
Lubricant capacity	¾ qt
Lubricant grade reference	par. 18

114. MAINTENANCE.

a. The transmission requires periodic checking of lubrication level. Mounting screws must be tight. Gearshift ball cap must be removed and cleaned as required. Keep all bond straps clean and securely tightened. Vent hole in transmission control housing must be kept clean at all times. Report transmission gear noise to higher authority.

115. REMOVAL.

a. Raise hood and fasten to windshield to prevent accidental closing. Open drain cock at bottom of radiator and drain cooling system. Remove radiator upper hose. Unscrew balls on shift levers. Remove bolts around transmission cover on floor, and remove cover. Remove transmission shift lever by unscrewing retainer collar at top of shift housing. Remove transfer case shift lever pin set screw. Remove lubricator in right end of shift shaft. Drive out shaft and remove levers. Place jack under engine oil pan. Remove exhaust pipe guard. Remove exhaust pipe clamp on skid plate. Remove skid plate. Remove bolts in front and rear propeller shaft universal joint flanges at transfer case end, and tie propeller shafts up to frame. Disconnect speedometer cable at transfer case. Remove transfer case support (snubber) bolt at cross member (fig. 73). Remove clevis pin in lower end of hand brake cable, and remove hand brake retracting spring. Remove clevis pin in clutch release cable yoke at cross tube lever. Disconnect engine stay cable at cross member. Remove bonding strap on transfer case and transmission. Unhook clutch pedal pull-back spring. Remove nuts on engine rear support insulator studs in cross member. Place a second jack under the transmission. Remove frame to cross member bolts at each end, and remove cross member. Place rope around the transmission. Push transmission to right, and remove clutch release lever tube from ball joint on transfer case. Remove four bolts holding transmission to bell housing. Remove two screws in inspection cover. Remove core and clutch release fork. Hold weight of assembly with rope, and remove jack from under transmission. Slide transmission assembly back until clutch shaft clears bell housing, lowering jack under engine just enough so that transmission will clear floor pan. Remove assembly from under vehicle.

b. Remove Transmission from Transfer Case. Clean outside of units. Drain lubricant from transmission and transfer case. Remove screws holding rear cover on transfer case, and remove cover and gasket. Remove cotter pin, nut, and washer on rear end of transmission main shaft. Pull off main shaft gear and oil slinger.

Figure 71—Gearshift Interlock Plunger

Remove four screws holding control housing on top of transmission, and remove housing. Remove shifter plate spring, and remove shifter plate. Loop piece of wire around main shaft to the rear of second speed gear, and attach wire tightly to front of transmission. Remove five screws holding transfer case to rear of transmission. Support transfer case, and tap lightly on end of transmission main shaft; at the same time draw transfer case away from transmission. NOTE: *Do not lose transmission gearshift interlock plunger* (fig. 71).

116. INSTALLATION.

a. **Assemble Transmission to Transfer Case.** Install transmission shift interlock plunger in position, using grease to hold it temporarily (fig. 71). Attach transfer case to transmission with five screws. Install shifter plate and spring. Install control housing on top of transmission with four screws. Using new gasket, install oil retaining washer on transmission main shaft in transfer case, with open face to rear. Install main drive gear with small power take-off gear to rear. Install washer, nut, and cotter pin. Install rear cover with screws and lock washers, using new gasket. Put lubricant in transmission and transfer case (par. 18).

b. **Install in Vehicle.** Place unit under vehicle. Place rope around transmission. Raise transmission, and insert transmission main drive gear shaft (clutch shaft) in hub of clutch plate and flywheel. Push transmission forward into position. Place jack under transmission to take the weight. Install clutch release fork on pivot ball, and install clutch release cable in end. Install transmission to bell housing bolts. Remove rope. Raise transmission with jacks. Push transmission to the right, and install clutch release tube on ball on transfer case. Install frame cross member. Remove jack under transmission. Install engine support insulator stud nuts at cross member. Place transfer case support rubber in place, and install bolt through cross member. Attach bond straps. Connect clutch release cable to lever. Connect hand brake cable and spring. Install engine stay cable and adjust so it is just taut. Install pedal pull-back spring. Adjust clutch release cable for pedal play (par. 109). Remove jack under engine. Connect speedometer cable at transfer case. Install transfer case shift levers and springs, driving shaft in from right side. Install shaft lock screw, and wire in place. Install lubricator in right end of shaft. Attach front and rear propeller shafts at transfer case. Install transmission skid plate. Install exhaust pipe clamp on skid plate. Install exhaust pipe guard. Install transmission shift lever in top of housing. Install transmission floor cover. Install gearshift lever balls. Install radiator upper hose. Fill cooling system, giving due attention to antifreeze, if required. Run engine until warm, and check coolant supply. Check cooling system for leaks. Lower hood and hook into place.

Section XXIII

TRANSFER CASE

Paragraph

117. DESCRIPTION AND DATA.

a. Description. The transfer case (fig. 72) is an auxiliary gear unit attached to the rear of the transmission. The transfer case is essentially a two-speed transmission which provides an additional

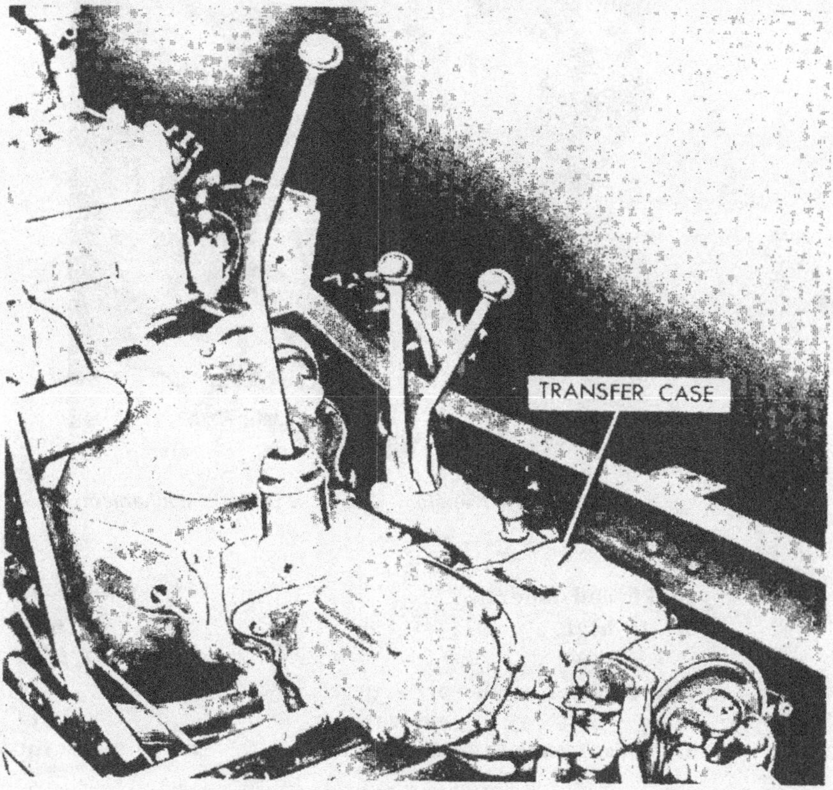

RA PD 305231

Figure 72—Transfer Case—Rear View

gear reduction for any selection of the transmission gears, also a means of engaging and disengaging power to drive the front axle. The shifting mechanism (fig. 73) is operated by two levers on the top of the case. A power take-off aperture is located at the rear just behind the transmission main shaft. The speedometer drive gear is located in the output shaft housing for the drive to the rear axle, and a hand brake is located on the same housing.

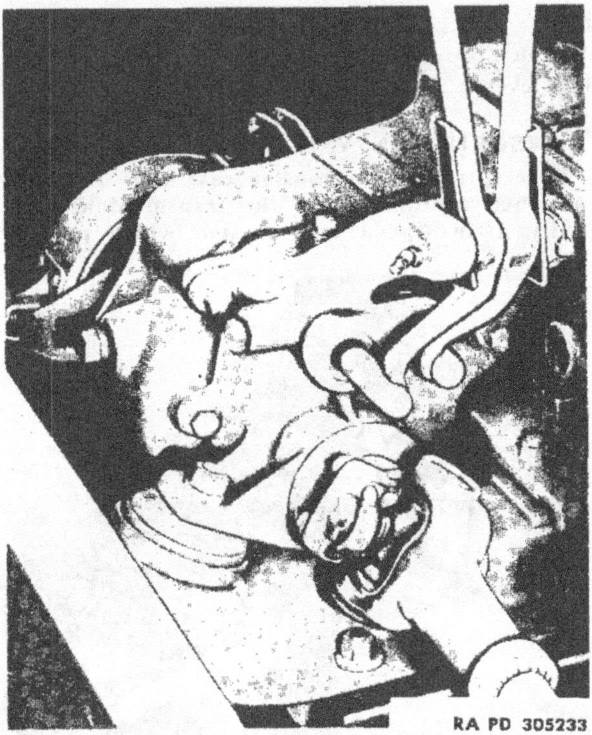

Figure 73—Transfer Case Shifting Mechanism

b. Data.

Make and model	Spicer—18
Ratio-high	1 to 1
-low	1.97 to 1
Speedometer teeth—drive gear	4
—driven gear	14
Lubricant capacity	1 1/2 qt

118. MAINTENANCE.

a. The transfer case requires lubrication at regular intervals. All mounting bolts must be kept tight. Universal joint yoke and companion flange nuts must be tight. Bond strap must be clean and tightened securely. Tighten drain and filler plugs. Report failure to stay in gear and noisy gears to higher authority.

119. REMOVAL.

a. Remove transmission and transfer case as an assembly as outlined in paragraph 115.

120. INSTALLATION.

a. Attach transfer case to transmission, and install in vehicle as outlined in paragraph 116.

121. SHIFTING LEVERS.

a. Removal. Unscrew accelerator footrest and remove 10 screws holding transmission floor cover, also remove two screws holding transfer case shift lever housing cover. Remove covers. Raise hood and secure to windshield. Remove transfer case shift lever set screw and lock wire. Unscrew hydraulic fitting from right end of shaft. Drive out shaft, and remove levers (fig. 73).

b. Installation. Place shift levers and springs in position. Drive in shaft. Install shaft set screw, and wire in place. Install hydraulic fitting in right end of shaft. Install transmission floor cover and accelerator footrest.

Section XXIV

PROPELLER SHAFTS AND UNIVERSAL JOINTS

122. DESCRIPTION AND DATA.

a. **Description.** Two propeller shafts are used, one to drive each axle. Two universal joints are used on each shaft (fig. 74). A splined slip joint is used at the rear of the front shaft, and, at the front of the rear shaft. NOTE: *The slip joint is marked with an arrow on the spline shaft and on the sleeve yoke, and these arrows must aline*

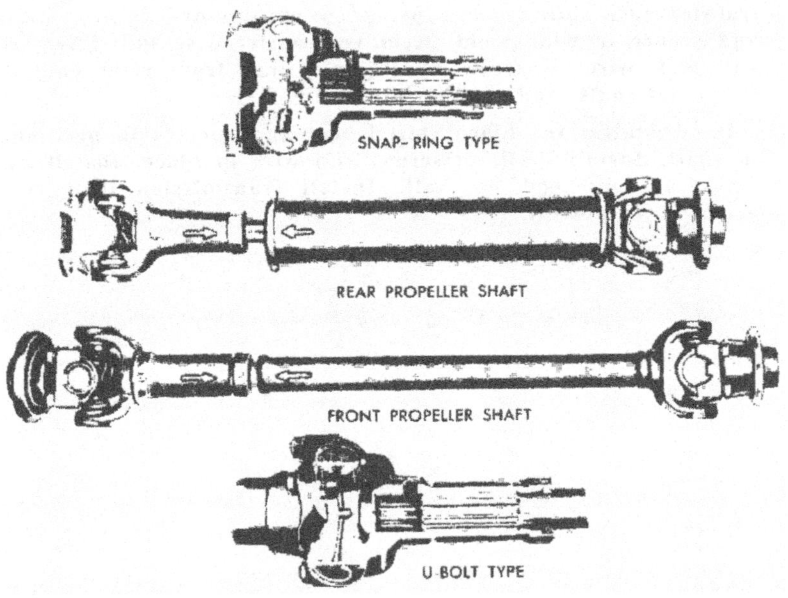

Figure 74—Propeller Shafts and Universal Joints

for correct assembly (fig. 74). The propeller shaft connecting the transfer case to the front axle has U-bolt type joints at both ends. The rear propeller shaft has a U-bolt type joint at the rear where it attaches to the rear axle, and a snap-ring type joint at the front end. The trunnion bearings are of the needle bearing type which are lubricated through a hydraulic fitting and an X-channel in the trunnion. Refer to lubrication instructions in paragraph 18.

b. **Data.**

Propeller shafts	
Make	Spicer
Installed length normal load-joint center to center	
Front	21 25/32 in.
Rear	21 5/8 in.
Front Universal Joint (front shaft)	
Type	Snap-ring and U-bolt
Model	1268
Rear Universal Joint (front shaft)	
Type	Snap-ring and U-bolt
Model	1261
Front Universal Joint (rear shaft)	
Type	Snap-ring
Model	1261
Rear Universal Joint (rear shaft)	
Type	Snap-ring and U-bolt
Model	1268

123. MAINTENANCE.

a. Propeller and universal joints require proper lubrication at regular intervals. Attaching bolts and nuts must be securely tightened. The yokes of front and rear universal joints must be assembled in the same plane. Snap rings must be securely locked in recess. Trunnion gaskets must be grease-tight. Tighten U-bolts evenly.

124. REMOVAL.

a. Front Propeller Shaft. Remove exhaust pipe shield. Remove U-bolts from universal joint yoke on axle. Remove U-bolts from universal joint yoke on transfer case. Remove shaft and universal joints as a unit.

b. Rear Propeller Shaft. Remove four bolts in front universal joint yoke flange. Remove two U-bolts from rear universal joint yoke on axle. Remove propeller shaft and universal joints as a unit.

125. INSTALLATION.

a. Front Propeller Shaft. Install propeller shaft and universal joint assembly in position on vehicle. Install U-bolts at axle. Install U-bolts at transfer case. NOTE: *Tighten U-bolts evenly.* Install exhaust pipe shield. Lubricate universal joints (par. 18 e).

b. Rear Propeller Shaft. Install propeller shaft and universal joint assembly in position. Install U-bolts in rear universal joint, and tighten evenly. Attach front universal joint flange at transfer case with four bolts. Lubricate universal joints (par. 18 c).

Section XXV

FRONT AXLE

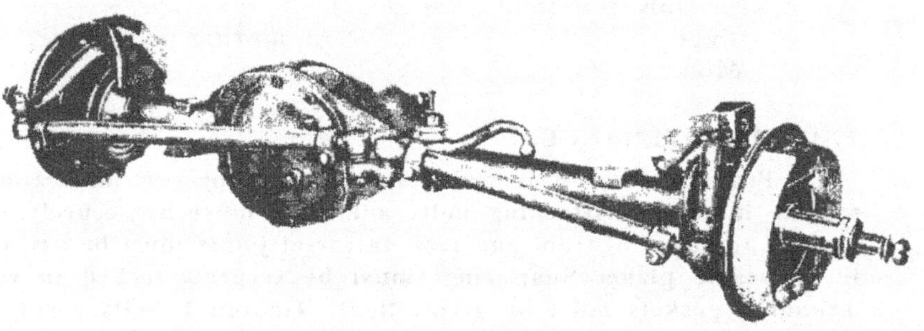

RA PD 305253

Figure 75—Front Axle

126. DESCRIPTION AND DATA.

a. Description. The front axle (fig. 75) is a full-floating type enclosing a front wheel driving unit having a single-reduction, two-pinion differential, and hypoid drive gears. The differential carrier housing is offset to the right so that the propeller shaft is located to the right of the engine for maximum ground clearance. A cover provides easy access to the differential unit. The front wheels are driven by axle shafts, each equipped with a constant-velocity type universal joint enclosed within a steering knuckle at the outer end of the axle housing. The differential assembly is the same as used in the rear axle. Power is transmitted by a propeller shaft from the transfer case, where a shift lever permits the vehicle operator to engage or disengage the drive.

b. Data.

Make and model . . . Spicer—25
Drive gear ratio . . . 4.88 to 1

¼-TON 4 x 4 TRUCK (WILLYS-OVERLAND MODEL MB and FORD MODEL GPW)

Drive	Hotchkiss (through springs)
Type	Full floating
Road clearance	8%$_{6}$ in.
Differential type	2 pinion
Differential drive gears	Hypoid
Differential bearings	Tapered roller
Turning angle	^{26} ^{de}6

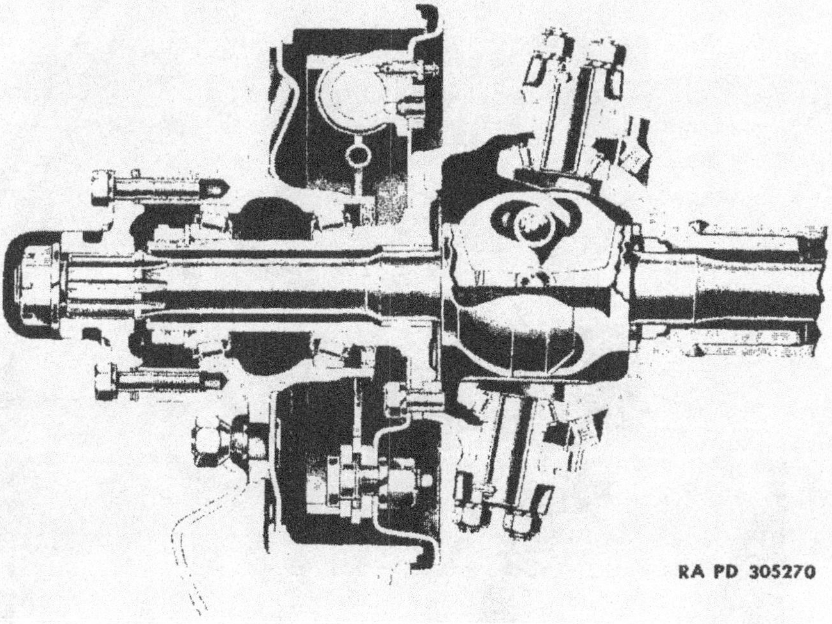

RA PD 305270

Figure 76—Front Wheel Hub

Tie rods:

Number	2
Right-hand length, center-to-center	24¼ in.
Left-hand length, center-to-center	17$^{11}/_{32}$ in.

Steering geometry:

King pin inclination	7½ deg
Wheel camber	1½ deg
Wheel caster	3 deg
Wheel toe-in	$^{3}/_{64}$ in. to $^{3}/_{32}$ in.

Bearings:

Differential side	Tapered roller
Pinion shaft	Tapered roller
Wheel hub	Tapered roller

Steering knuckle	Tapered roller
Steering bell crank	Tapered roller
Lubricant capacity	$1\frac{1}{4}$ qt

127. MAINTENANCE.

a. Correct any lubricant leakage. Lubricate differential and steering housings, wheel bearings, and steering control as required (par.

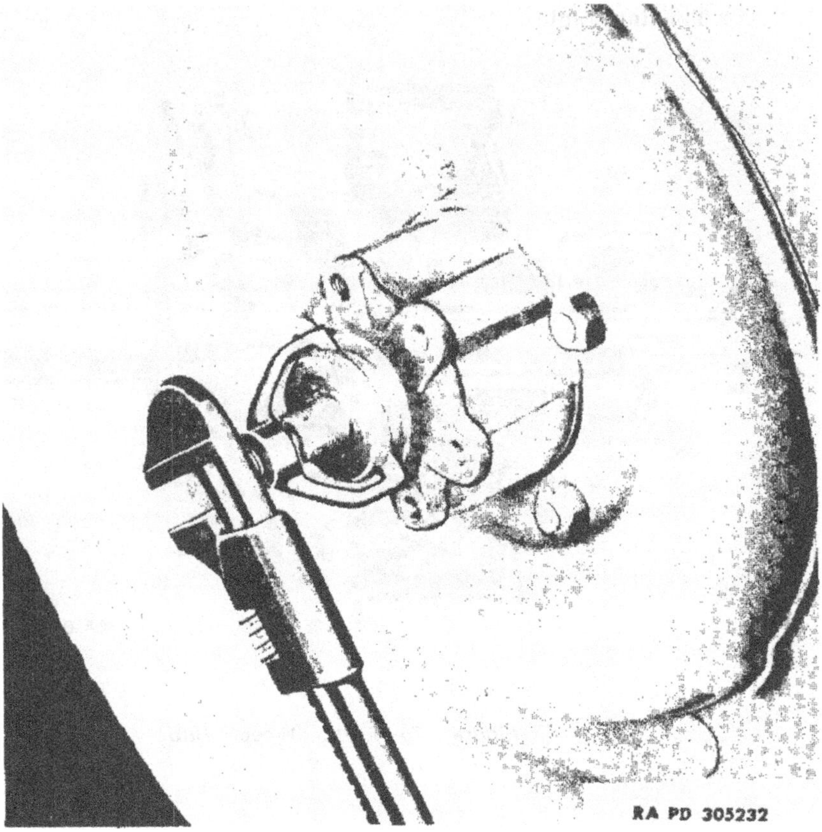

Figure 77—Removing Driving Flange, Using Puller (41-P-2905-60)

18). Keep vent cleared of dirt. Wheel bearings must be properly adjusted (par. 128). Replace damaged brake drums and hubs, and steering control rods (par. 133). Check wheel toe-in periodically, and correct if necessary (par. 135). Keep all mounting bolts tight. Report to higher authority any caster and camber trouble, or unusual noises.

128. WHEEL BEARINGS.

a. Adjustment. Raise front of vehicle so that tire clears floor. Pry off hub cap (fig. 76). Remove axle shaft nut cotter pin, nut, and washer. Remove driving flange screws. With puller, pull off flange (fig. 77). NOTE: *Do not lose flange shims.* Bend lip of nut lock washer away from nut. Remove lock nut with box-type socket wrench (fig. 78). Remove lock washer. Spin wheel and tighten wheel bearing nut until wheel binds. Back off nut about one-sixth turn, or more if necessary, until wheel turns freely. Install lock washer and lock nut. NOTE: *Bend over lip of lock washer against lock nut.* Check adjustment of bearings by gripping front and rear side of tire and moving it from side to side. A slight perceptible shake should be felt in the bearings. Install flange shims and flange. Check axle shaft end play by tightening flange nut without the lock washer. Swing wheel to maximum left or right; have punch mark on end of axle shaft up or down. Back off nut until 0.050-inch thickness gage will go between hub and nut. Shaft will move in by the amount of end play when end is tapped with soft hammer. Measure clearance between nut and flange, and deduct amount from 0.050 inch to determine end play. If less than 0.015 inch or more than 0.035 inch, correct thickness of shim pack (with Rzeppa joint disregard these instructions and use 0.060-inch shim pack). Install axle shaft lock washer, nut, and cotter pin. Install new hub cap.

b. Removal. Loosen wheel stud nuts. NOTE: *Wheel studs have left-hand threads on left side of vehicle.* Raise front of vehicle so that tire clears floor. Remove wheel stud nuts and remove wheels. Pry off hub cap. Remove axle shaft nut cotter pin, nut, and washer. Remove driving flange screws and using puller, pull flange. NOTE: *Do not lose flange shims.* Bend lip of nut lock washer away from lock nut and remove nut. Remove lock washer. Remove wheel bearing nut and bearing lock washer. Shake wheel until outer bearing comes free of hub, and lift off wheel. Drive or press out inner bearing along with oil seal. Turn wheel over, and drive or press out outer bearing cup. Clean lubricant out of hub, and wash all parts in dry-cleaning solvent.

e. Installation. Press bearing cups solidly into place in hub. Spread Vitj-inch layer of lubricant inside hub to prevent rust. Thoroughly lubricate inner bearing cone and roller assembly. NOTE: *Pack lubricant thoroughly into bearing rollers and cage.* Install bearing in hub. Press oil seal into hub (with lip of seal toward bearing), until seal is even with end of hub. NOTE: *Before installation, soak seal in oil to soften leather.* Lubricate outer bearing cone and roller assembly. Install wheel on axle. Install outer bearing, lock washer, and nut. Adjust wheel bearings, and complete installation of parts (subpar. a above).

129. WHEEL GREASE RETAINER.

a. Removal. Remove retainer as outlined in paragraph 128 b.

b. Installation. Install retainer as outlined in paragraph 128 c.

130. WHEEL HUB.

a. Removal. Remove wheel and hub (par. 128 b). Support braice drum inside at hub, and drive out studs. Remove brake drum.

b. Installation. Assemble brake drum on hub. Install new wheel studs. NOTE: *Left-hand thread studs are used in wheels for left side of vehicle.* Support studs and swedge shoulder over against tapered hole in hub. Install hub on axle and mount wheel (par. 128 c).

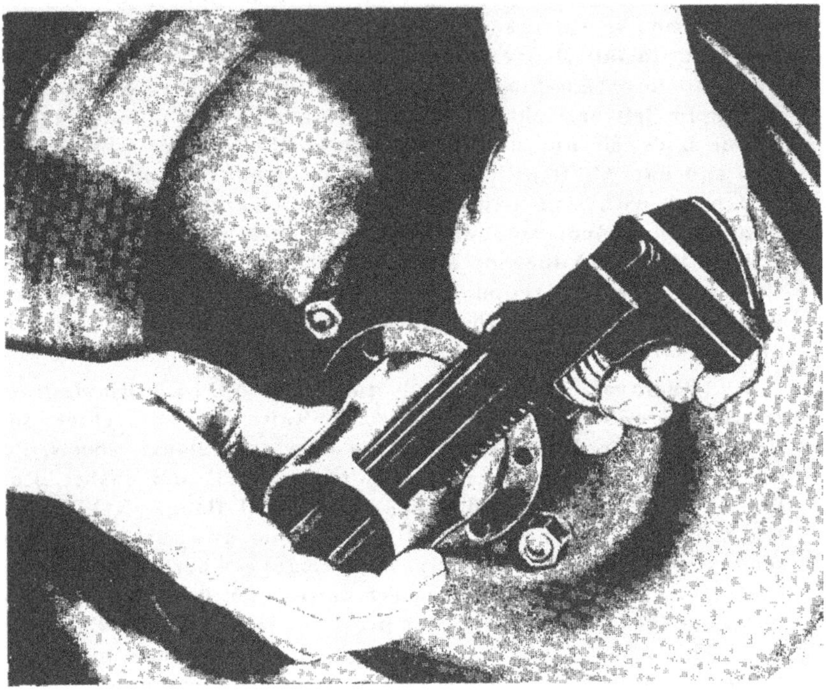

RA PD 305235

Figure 78—Removing Lock Nut, Using Wrench (41-W-3825-200)

131. BRAKE DRUMS.

a. Removal. Remove wheel hub and drum as outlined in paragraph 130 a.

b. Installation. Install drum and wheel hub as outlined in paragraph 130 b. Adjust brakes (par. 148).

132. STEERING KNUCKLE HOUSING OIL SEAL.

a. Removal. Raise front of vehicle. Remove screws holding oil seal in place, and remove both halves of oil seal assembly (fig. 79).

b. Installation. NOTE: *Before installing new oil seal smooth spherical surface of axle with aluminum oxide abrasive cloth.* Grease spherical surface of axle, also oil seal. Install seal in place so that ends fit snugly together, and tighten in place. Check lubricant level in steering knuckle housing, and replenish if necessary (par. 18).

133. STEERING TIE ROD.

a. Removal. Remove tie rod cotter pins and nuts from tie rod ends (fig. 75). Drive out tie rod ends from steering arms, and remove dust washers and springs.

b. Installation. Install dust washers and springs on tie rod ends. Install tie rod ends in steering knuckle arm and bell crank, and secure with nuts and cotter pins. Check wheel alinement, and adjust if necessary (par. 135).

RA PD 305240

figure 79—Steering Knuckle Oil Seal

134. DRAG LINK BELL CRANK.

a. Removal. Remove cotter pin in front end of steering connecting rod. Remove slotted adjusting plug (ball seat). Lift rod off bell crank ball. Remove cotter pins and nuts on tie rod ends at bell crank. Drive tie rod ends out of bell crank. NOTE: *Do not lose dust washers and springs.* Remove cotter pin in bell crank stud and remove nut, dust washer, and thrust washer. Remove bell crank. Clean all parts in dry-cleaning solvent. To remove bell crank stud, remove thrust washer, and drive out tapered lock pin toward left front wheel. Drive stud up out of axle.

b. Installation. If bell crank has been removed, drive stud into axle so that slot will line up with tapered pin hole. Drive tapered pin into position, and stake edge of hole at large end. Install thrust washer on stud. Lubricate roller bearings, and install on stud. Install thrust washer, dust washer, nut, and cotter pin. Install steering connecting rod. Install tie rod ends in bell crank arm, and secure each with nut and cotter pin. Check front wheel toe-in, and adjust if necessary (par. 135).

135. WHEEL ALINEMENT (TOE-IN).

a. **Caster and Camber.** Caster is the backward tilt of the axle. Camber is the outward tilt of the wheels at the top. If these conditions require attention, notify higher authority.

Figure 80—Brake Hose

b. **Toe-in.** Wheel toe-in is the difference in distance between the front wheels at the front and at the rear of the axle. To adjust toe-in, set tie rod arm of steering bell crank at right angles to front axle. Use straightedge or line against outside of left wheels, as a guide. Adjust left tie rod so that left wheel is straight ahead. While bell crank remains at right angle to axle, check right front wheel, and adjust tie rod if necessary. Set toe-in of front wheels at 3/64 inch to 3/32 inc by shortening right tie rod approximately one turn.

136. REMOVAL.

a. Loosen wheel stud nuts. Raise front of vehicle, and support underframe side members at rear of spring pivot brackets. Remove wheels. Disconnect brake line at front cross member (fig. 80). Remove universal joint U-bolts at front axle. Jack up front springs. Remove axle spring clip nuts and clips. Remove spring pivot bolt at rear end of right spring. Remove jacks from under springs. Disconnect steering connecting rod at bell crank. Install a jack between left spring and frame. Spread spring until axle assembly will clear. Move axle assembly to the right, and remove. Remove brake hose from axle.

137. INSTALLATION.

a. Attach brake hose at axle. Install axle assembly on springs. Remove jack from between left spring and frame. Install right spring pivot bolt. Position axle on springs. Jack up springs, and install spring clips, plates, and nuts. Remove jacks from under springs. Connect brake hose at cross member. Install dust cover on bell crank, and attach steering connecting rod. Attach propeller shaft. Draw universal joint U-bolts up evenly. Lubricate front axle universal joints, and check axle lubricant (par. 18). Adjust brakes if necessary (par. 147). Remove master cylinder inspection cover on toeboard between foot pedals. Fill master cylinder, and bleed brakes (par. 151). Replace master cylinder inspection cover. Install wheels and adjust (par. 128 a). Lower vehicle to floor.

Section XXVI

REAR AXLE

Paragraph

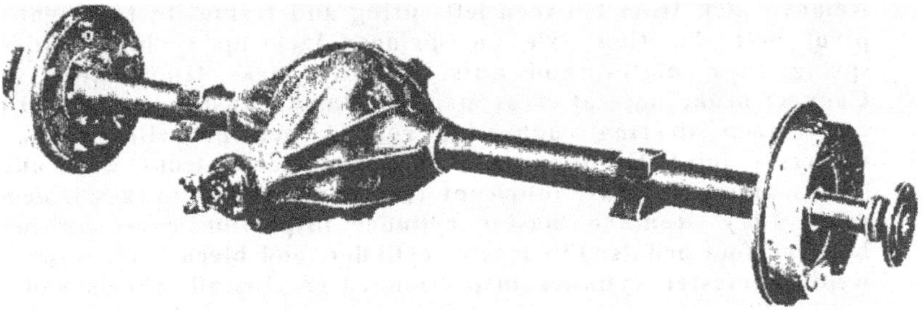

RA PD 305254

Figure 81—Rear Axle

a. **Description.** The rear axle (fig. 81) is a full-floating type enclosing a single-reduction driving unit, two-pinion differential, and hypoid-drive gears. The differential carrier housing is offset to the right so that the propeller shaft will have a straight drive from the transfer case. A cover provides easy access to the differential unit. The axle shafts are splined to fit into the differential side gears, and flanged at the outer end where they attach to the wheel hub. The wheel bearings are adjusted by two nuts threaded onto the axle tube.

b. **Data.**

Make and model	Spicer 23-2
Drive gear ratio	4.88 to 1
Drive	Hotchkiss (through springs)
Type	Full floating
Road clearance	8$\frac{7}{16}$ in.
Differential type	Two-pinion
Differential bearings	Tapered roller

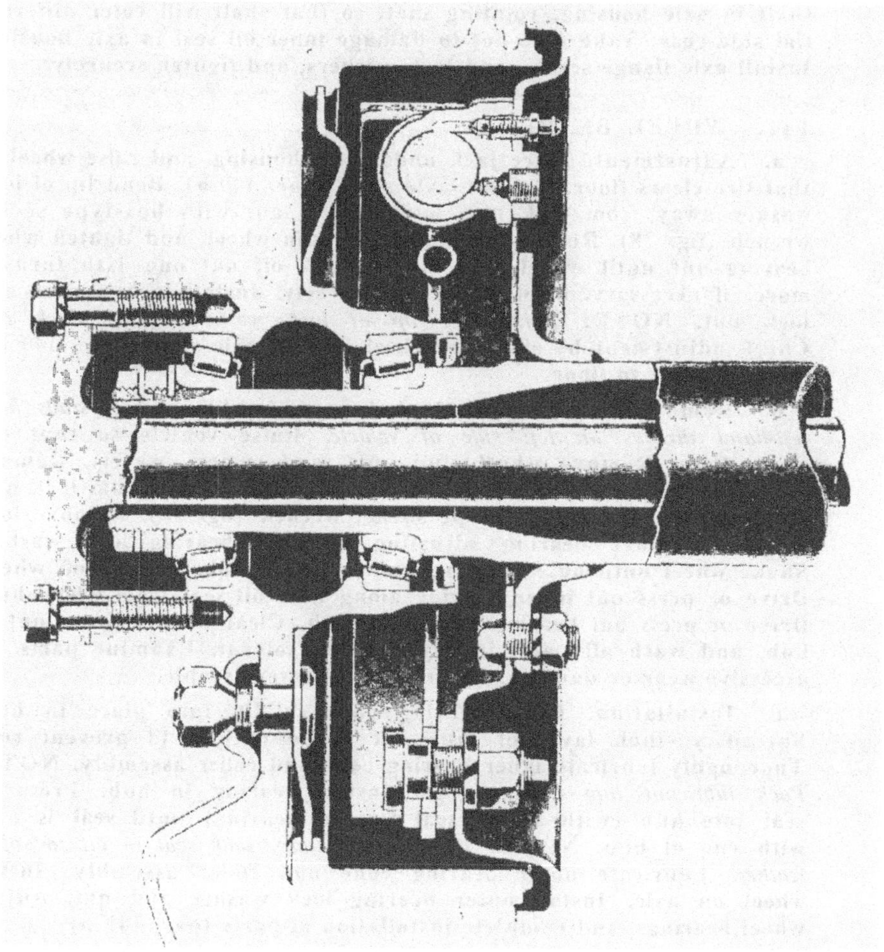

Figure 82-Rear *Wheel* Hub

139. MAINTENANCE.

a. Correct any lubricant leakage. Lubricate differential and wheel bearings as required (par. 18). Keep vent cleared of dirt. Wheel bearings must be properly adjusted (par. 141). Replace damaged drums and hubs (par. 143). Keep all mounting bolts tight. Report unusual noise to higher authority.

140. AXLE SHAFT.

a. Removal. Remove six axle shaft flange screws and lock washers. Pull out axle shaft (fig. 82), and remove flange gasket.

Ij. Installation. Install new axle shaft flange gasket. Install axle shaft in axle housing, rotating shaft so that shaft will enter differential side gear. Take care not to damage inner oil seal in axle housing. Install axle flange screws and lock washers, and tighten securely.

141. WHEEL BEARINGS.

a. Adjustment. Place jack under axle housing, and raise wheel so that tire clears floor. Remove axle shaft (par. 140 a). Bend lip of lock washer away from lock nut, and remove nut with box-type socket wrench (fig. 78). Remove lock washer. Spin wheel, and tighten wheel bearing nut until wheel just binds. Back off nut one-sixth turn or more, if necessary, until wheel turns freely. Install lock washer and lock nut. NOTE: *Bend over lip of lock washer against lock nut.* Check adjustment by shake of wheel. Install axle shaft (par. 140 b). Lower vehicle to floor.

b. Removal. Loosen wheel stud nuts. NOTE: *Wheel studs have left-hand threads on left side of vehicle.* Raise vehicle so that tire clears floor. Remove wheel stud nuts, and remove wheels. Remove axle shaft (par. 140 a). Bend lip of lock washer away from lock nut, and remove nut with box-type socket wrench (fig. 78). Remove lock washer. Remove bearing adjusting nut and bearing lock washer. Shake wheel until outer bearing comes free of hub, and lift off wheel. Drive or press out inner bearing along with oil seal from wheel hub. Drive or press out bearing cups from hub. Clean old lubricant out of hub, and wash all parts in dry-cleaning solvent. Examine parts for excessive wear or damage, and replace if unserviceable.

c. Installation. Press bearing cups solidly into place in hub. Spread y^-inch layer of lubricant inside of hub to prevent rust. Thoroughly lubricate inner bearing cone and roller assembly. NOTE: *Pack lubricant into bearing cage.* Install bearing in hub. Press oil seal into hub (with lip of seal toward bearing) until seal is even with end of hub. NOTE: *Before installation, soak seal in oil to soften leather.* Lubricate outer bearing cone and roller assembly. Install wheel on axle. Install outer bearing lock washer and nut. Adjust wheel bearings, and complete installation of parts (par. 141 a).

142. WHEEL BEARING GREASE RETAINER.

a. Removal. Remove retainer as outlined in paragraph 141 b.

b. Installation. Install retainer as outlined in paragraph 141 c.

143. WHEEL HUB.

a. Removal. Remove wheel and hub as outlined in paragraph 141 b. To remove brake drum from hub, support brake drum at hub, and drive out studs.

b. Installation. Place brake drum on hub. Install new wheel studs. NOTE: *Left-hand thread studs are used in wheels on left side of vehicle.* Support studs and swedge shoulder over against tapered hole in hub. Install hub on axle and mount wheel (par. 141 c). Tighten wheel stud nuts securely. Check brake action.

144. BRAKE DRUM.

a. Removal. Remove wheel hub and drum as outlined in paragraph 143 a.

b. Installation. Install drum and wheel hub as outlined in paragraph 143 b.

145. REAR AXLE REPLACEMENT.

a. Loosen wheel stud nuts. Raise rear of vehicle and support underframe side member just ahead of spring pivot brackets. Remove wheels. Remove universal joint U-bolts at rear axle. Disconnect brake hose at frame cross member. Remove brake hose at axle. Place jack under each rear spring. Remove spring clip nuts, clips, and plates. Remove jacks from under Springs, place between frame and spring, and spread spring. Remove axle, sliding it to left until right end clears spring, then slide to right and remove.

b. Installation. Install axle assembly on springs. Remove jacks from between springs, and place under each spring. Position axle on springs. Install spring clips, plates, lock washers, and nuts. Tighten nuts securely. Remove jacks from under springs. Attach propeller shaft. Draw universal joint U-bolts up evenly. Attach brake hose at axle, then at frame cross member. Check axle lubricant. Remove master cylinder inspection cover on toeboard between foot pedals. Fill master cylinder, and bleed brakes (par. 151). Replace master cylinder inspection cover. Adjust brakes if necessary (par. 148). Install wheels. Lower vehicle to floor.

Section XXVII

BRAKES

146. DESCRIPTION AND DATA.

a. Description. The service, or foot brake, system is of the hydraulic type with brakes in all four wheels (fig. 83). The parking, or hand brake, is cable-controlled and mounted on the rear side of the transfer case (fig. 84). The service, or foot brakes, are of the two-shoe, double-anchor type. The brake pedal, through a connection, operates a piston in the master cylinder to force brake fluid through the lines to the brake cylinders in the wheels. The fluid enters the wheel cylinders between two pistons of equal diameter, forcing them apart to apply the brake shoes against the drums. Releasing the brake pedal permits the brake fluid to flow back through the lines to the master cylinder. Adjustments are provided to compensate for wear of the brake linings. The hand brake is designed for parking the vehicle, or as an emergency brake. The hand brake lever is located at the center of the instrument panel. Pulling out on the lever draws a flexible cable through a conduit to actuate an external contracting brake band at the rear of the transfer case. The brake cable is of a predetermined length, and cannot be adjusted. When adjustment is required, the brake band lining will be worn to the point where replacement is necessary. Adjustments are provided on the brake to set the band correctly, and to limit the release action.

b. Data.

Service brakes:

Type .... Four-wheel, hydraulic
Size .... 9 in. x 1¾ in.
Fluid capacity .... ¼ qt

Master cylinder:

Type .... Combination reservoir and cylinder
Size .... 1 inch

Wheel cylinders:

Type .... Straight bore
Size .... Front, 1 in.; rear, ¾ in.

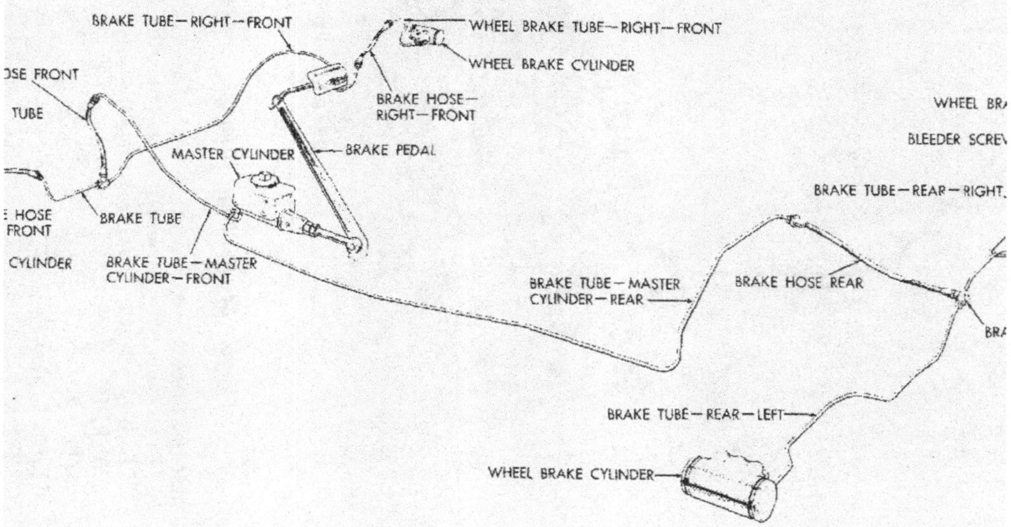

Figure 83—Service (Foot) Brake System

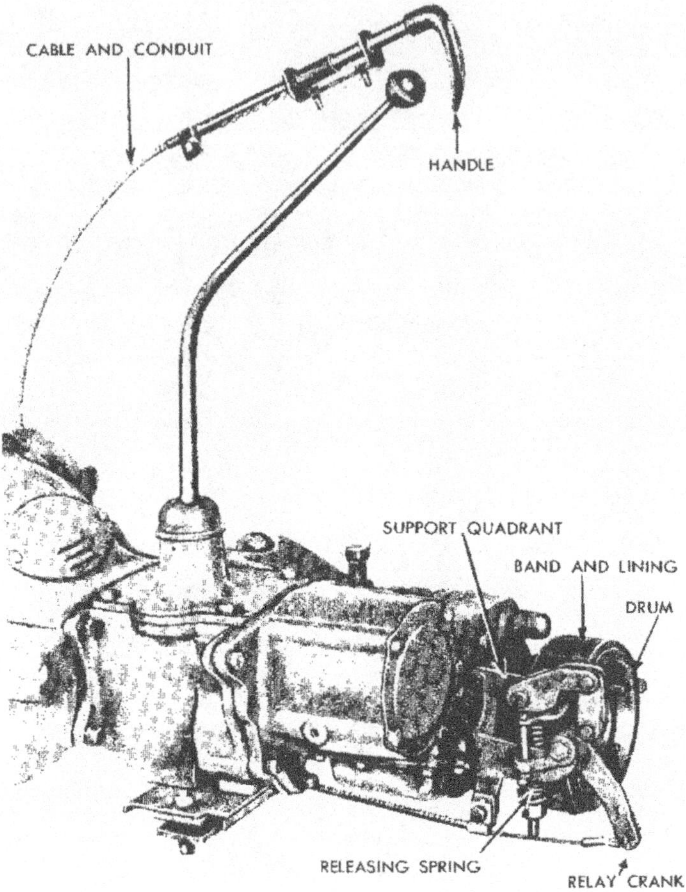

RA PD 305256

Figure 84—Parking (Hand) Brake System

Brake shoes:	
Lining length—forward shoe (moulded)	10 7/32 in.
Lining length—reverse shoe (moulded)	6 39/64 in.
Width	1 3/4 in.
Thickness	3/16 in.
Hand brake:	
Type	Mechanical
Lining length (woven)	18 9/16 in.
Width	2 in.
Thickness	5/32 in.

147. MAINTENANCE AND ADJUSTMENT.

a. The service, or foot, brakes require periodic checking of the brake fluid supply in the master cylinder. Keep master cylinder supplied with fluid to avoid air entering the lines. Wheel bearings and brakes must be properly adjusted to provide emergency stops. All brake lines, hoses, and connections must be tight and leakproof. Scored brake drums or saturated brake linings must be replaced. Clean brake drums when wheels are removed. Brake anchor bolt and eccentric adjustment bolt lock nuts must be kept tight. Brake backing plate screws and axle spring clips must be kept tight. Brake pedal must have 1/2 inch free travel to assure full release of brakes. Brake control linkage must be free to operate, and should be inspected periodically for condition.

RA PD 305260

Figure 85 — Wheel Brake

b. **Adjustment.** Adjust brake pedal free travel by lengthening or shortening brake master cylinder eyebolt so that pedal has 1/2 inch free play (par. 148). Follow procedure outlined in paragraph 148 to adjust brakes when lining has worn so that brake pedal goes almost to the toeboard. Three adjustments are provided on the hand brake (par. 152).

148. SERVICE (FOOT) BRAKES.

a. **Adjustment (minor).** Adjust brake pedal free play to one-half inch by lengthening or shortening brake master cylinder eyebolt. Set lock nut securely. Raise vehicle until tires clear floor. NOTE: *Do not adjust brakes when drums are hot.* Loosen eccentric lock nut on forward shoe of one brake (fig. 86). Place wrench on eccentric so

Figure 86—Wheel Brake Adjustment Points

that handle extends up. Rotate wheel, and turn wrench handle toward wheel rim, or forward, until brake drags. Turn wrench in opposite direction until wheel turns freely. Hold wrench on eccentric, and tighten lock nut. Loosen eccentric lock nut on reverse dhoe. Place wrench on eccentric with handle up. Rotate wheel, and turn wrench toward wheel rim, or to the rear, until brake drags. Turn wrench in opposite direction until wheel turns freely. Hold wrench on eccentric, and tighten lock nut. Make the same adjustment on the other wheel brakes. Replenish brake fluid in master cylinder (par. 149). Lower vehicle to floor. Apply brake pedal to test brakes.

h. Adjustment (major). Adjust brake pedal free play to one-half inch by lengthening or shortening brake master cylinder eye-bolt. Set lock nut securely. Raise vehicle until tires clear floor. NOTE: *Do not adjust brakes when drums are hot.* Remove wheel stud nuts, and remove wheels from hubs. Insert 0.008-inch thickness gage through slot in brake drum, and turn drum so that gage is at upper (toe) end of forward brake lining. NOTE: *Check clearance 1 inch from end of lining.* Loosen eccentric lock nut on forward brake shoe. Place wrench on eccentiic so that handle is up, and turn wrench handle toward wheel rim, or forward, until 0.008-inch clearance is obtained by feel of gage. Hold wrench on eccentric, and tighten lock nut. Turn brake drum so that gage is at upper end of reverse brake shoe lining. Loosen eccentric lock nut on reverse shoe. Place wrench on eccentric so that handle is up, and turn wrench handle toward wheel rim, or to the rear, until 0.008-inch clearance is obtained by feel of gage. Hold wrench on eccentric, and tighten lock nut. Remove 0.008-inch thickness gage, and insert 0.005-inch gage in slot. Turn brake drum so that gage is at lower (heel) end of forward brake shoe

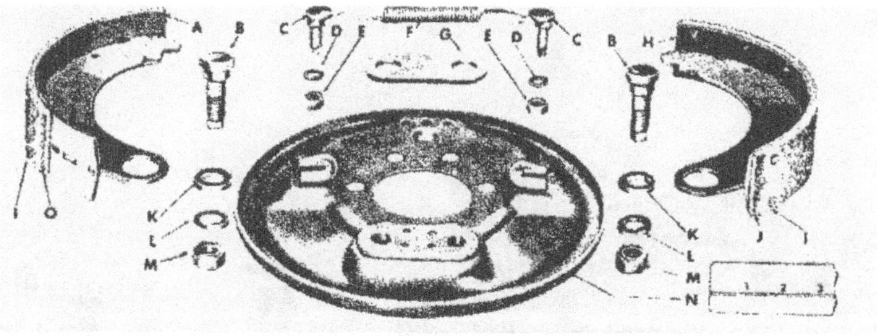

A SHOE AND LINING ASSEMBLY-REVERSE
B ANCHOR PIN
C ECCENTRIC
D ECCENTRIC LOCK WASHER
E ECCENTRIC NUT
t RETURN SPRING
0 ANCHOR PIN PLATE

H SHOE AND LINING ASSEMBLY—FORWARD
I LINING TUBULAR BRASS RIVET
J LINING—FORWARD
K ANCHOR PIN CAM
L ANCHOR PIN LOCK WASHER
M ANCHOR PIN NUT
N BACKING PLATE ASSEMBLY
O LINING—REVERSE

RA M> 30527>

Figure 87—Wheel Brake Shoes, *Disassembled*

lining. Loosen lock nut on anchor pin of forward shoe. Place wrench on anchor pin with handle down, and punch marks on ends of anchor pins toward each other; turn wrench toward rim, or forward, until 0.005-inch clearance is obtained by feel of gage. Hold anchor pin and tighten lock nut. Turn brake drum so gage is at lower end of reverse brake shoe lining. Loosen anchor pin lock nut on reverse shoe. Place wrench on anchor pin with handle down, and punch mark on end of anchor pin toward other anchor pin; turn wrench handle toward rim, or to the rear, until 0.005-inch clearance is obtained by feel of gage. Hold anchor pin and tighten lock nut. Follow same procedure on the other three brakes. Check amount of fluid in master cylinder (par.

149), and apply foot brake pedal to test brakes. Bleed brakes if *soft* pedal is experienced (par. 151). Install wheel. Lower vehicle to floor.

c. **Removal of Brake Shoes and Linings.** Raise vehicle. Remove wheel hubs (pars. 128 and 141). Loosen eccentric lock nuts (fig. 87). Turn eccentric so that low side is against the shoes. Install brake cylinder clamp to hold pistons in place. Remove brake shoe return spring. Remove anchor pin nuts, lock washers, anchor pins, and anchor pin plate from backing plate. Remove brake shoes. Remove brake shoe anchor pin cam. Inspect exterior of wheel brake cylinder for leakage of brake fluid. If leakage is apparent, replace cylinder assembly (par. 150).

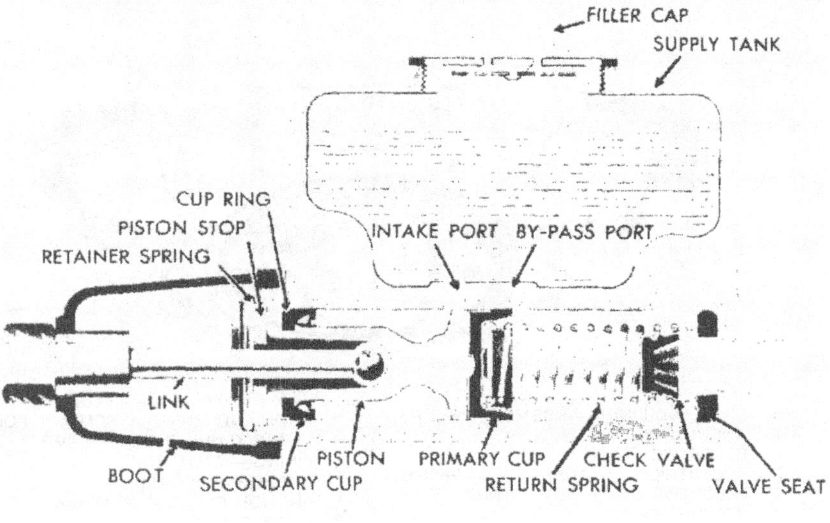

Figure 88—Master Cylinder

d. **Installation of Brake Shoes and Linings.** Install cam in brake shoes. Install anchor pin plate on anchor pins; install pins in brake shoes, and mount assembly on brake backing plate. NOTE: *Forward shoe has longest lining.* Install brake return spring, and remove brake cylinder clamp. Install brake anchor pin lock washers and nuts. NOTE: *Turn brake anchor pins so that punch marks on ends are toward each other. Do not tighten anchor nuts.* Install hubs (pars. 128 and 141). Make major brake adjustment (par. 149 b).

149. MASTER CYLINDER.

a. **Removal.** Raise hood and disconnect battery ground at battery terminal. Remove two bolts holding master cylinder shield and remove shield. Pull stop light switch wires out of terminal on switch. Remove stop light switch. Remove outlet fitting screw. Remove master cylinder front screw attaching cylinder to frame. Remove master cylinder rear bolt nut. Remove cotter pin holding master cylinder tie bar on pedal cross shaft. Remove master cylinder boot (fig. 88). Remove master cylinder and tie bar. Remove tie bar from master cylinder.

b. Installation. Fill master cylinder with brake fluid. Install tie bar and rear bolt on master cylinder, and install master cylinder in frame with tie bar on pedal shaft. Install cotter pin in pedal shaft. Install eyebolt link in cylinder. Install master cylinder front screw,

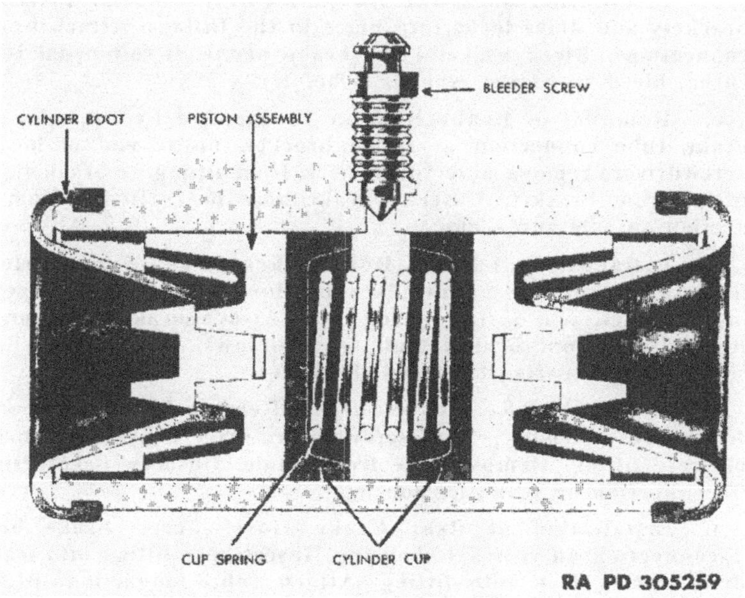

Figure 89—Wheel Cylinder

and tighten rear bolt. Install master cylinder boot with drain hole down. Install outlet fitting bolt. Install stop light switch. Insert stop light wires in terminals.. Install master cylinder shield with two bolts. Bleed brakes (par. 151). Attach battery ground cable. Lower hood and hook.

150. WHEEL CYLINDER.

a. Removal. Raise vehicle so that tire clears floor. Remove wheel and hub (pars. 128 and 141). Remove brake shoe return spring. Spread shoes until clear of brake cylinder. Disconnect brake tube at backing plate. Remove two screws holding cylinder to backing plate, and remove cylinder.

b. Installation. Place cylinder in position on backing plate, and attach with two screws and lock washers. Attach brake tube. Enter brake shoes in slots of cylinder pistons (fig. 89). Install brake shoe return spring. Replace wheel and hub (pars. 128 and 141). Bleed brake (par. 151). Apply foot brake pedal to test brakes. If soft pedal is experienced, bleed all brakes. Lower vehicle to floor.

151. FLEXIBLE LINES, HOSES, AND CONNECTIONS.

a. Removal of Brake Hose at Front Wheels. Remove brake tube connections at each end. With screwdriver slip hose lock off ends of hose fitting, and remove hose.

b. Installation of Brake Hose at Front Wheels. Place hose in brackets and drive locks into place in the fittings. Attach brake tube connections. Bleed brake. Press brake pedal; if soft pedal is experienced, bleed all brakes (subpar. s below).

e. Removal of Brake Hose at Frame and Front Axle. Remove brake tube connection at frame bracket, upper end of hose. With screwdriver, remove hose spring lock from fitting at bracket. Remove fitting from bracket. Unscrew brake hose lower fitting from T-connection on axle and remove.

d. Installation of Brake Hose at Frame and Front Axle. Screw brake hose lower fitting into T-connection on axle. Insert upper fitting into bracket, and install spring lock. Attach brake tube connection. Bleed both front brakes (subpar. s below). Press brake pedal; if soft pedal is experienced, bleed all brakes.

e. Removal of Rear Brake Hose. Remove brake tube connection frame cross member. With screwdriver, drive brake hose spring lock off hose fitting. Remove hose from frame. Unscrew hose fitting from T-connection on rear axle housing.

f. Installation of Rear Brake Hose. Screw brake hose into T-connection on rear axle housing. Insert hose fitting into frame, and drive spring lock into fitting. Attach tube connection. Bleed both rear brakes (subpar. s below). Press brake pedal; if soft pedal is experienced, bleed all brakes.

g. Removal of Master Cylinder to Front Hose Brake Tube. Remove clip from frame. Disconnect tube from brake hose fitting (frame to axle). Disconnect tube from master cylinder connection, and remove tube.

h. Installation of Master Cylinder to Front Hose Brake Tube. Connect tube at master cylinder. Connect tube at brake hose (frame to front axle). Install tube clip at frame. Bleed front brakes (subpar. s below). Press brake pedal; if soft pedal is experienced, bleed all brakes.

i. Removal of Master Cylinder to Rear Hose Brake Tube. Remove clip on underside of frame rear cross member. Remove clip on frame side member. Disconnect tube at rear brake hose. Remove master cylinder shield. Disconnect tube at master cylinder. Withdraw tube to rear of vehicle.

j. Installation of Master Cylinder to Rear Hose Brake Tube. Install tube in frame side member. Connect tube to master cylinder, and install master cylinder shield. Install tube in frame rear cross member, and attach to hose fitting. Install tube clips on frame side member and rear cross member. Bleed rear brakes (subpar. s below). Press brake pedal; if soft pedal is experienced, bleed all brakes.

k. Removal of Tee to Front Hose Brake Tube—Left. Disconnect brake tube at tee connection. Disconnect tube at brake hose fitting and remove tube.

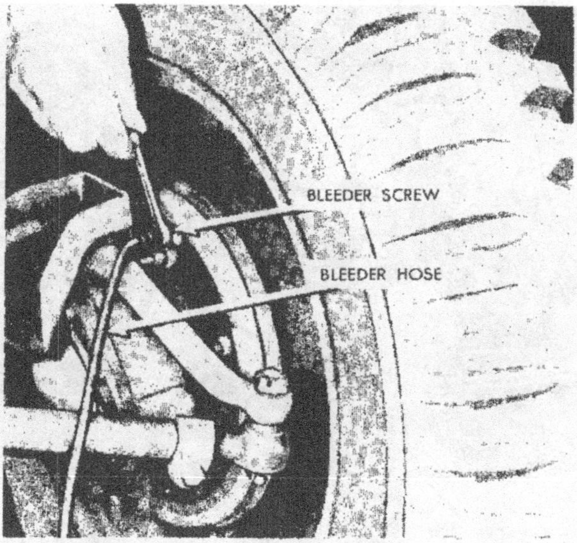

RA PD 305263

Figure 90—Bleeding Brakes

l. Installation of Tee to Front Hose Brake Tube—Left. Connect brake tube at tee connection. Connect tube at hose fitting. Bleed left brake (subpar. s below). Press brake pedal; if soft pedal is experienced, bleed all brakes.

m. Removal of Tee to Front Hose Brake Tube—Right. Remove clips and clamps on axle. Disconnect tube at tee connection. Disconnect tube at hose fitting and remove tube.

n. Installation of Tee to Front Hose Brake Tube—Right. Connect brake tube at tee connection. Connect tube at brake hose fitting. Install clips and clamps on axle. Bleed right front brake (sub-

all brakes.

o. Removal of Front Wheel Cylinder to Hose Brake Tube. Disconnect tube at brake hose. Disconnect tube at wheel cylinder and remove tube.

p. Installation of Front Wheel Cylinder to Hose Brake Tube. Attach brake to wheel cylinder. Attach tube to hose fitting. Bleed brake (subpar. s below). Press brake pedal; if soft pedal is experienced, bleed all brakes.

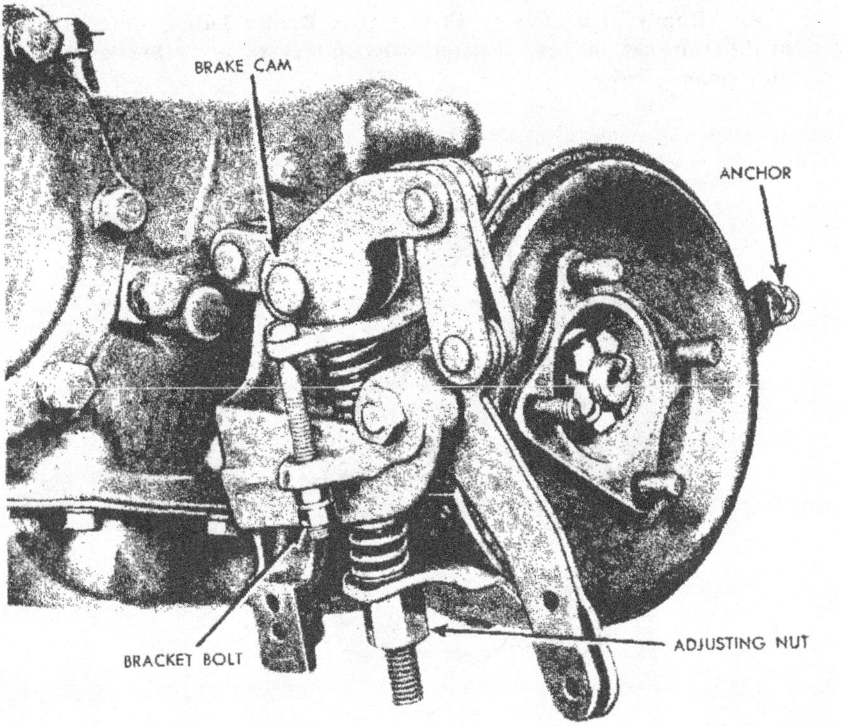

Figure 91—Parking (Hand) Brake

q. Removal of Tee to Rear Brake Tube—Right. Disconnect brake tube at tee connection. Disconnect tube at wheel cylinder. Remove tube clamp on axle. Bend tube slightly and remove.

r. Installation of Tee to Rear Brake Tube. Attach brake tube to wheel cylinder. Attach tube to tee connection. Install clamp on axle. Bleed brake (subpar. s below). Press brake pedal; if soft pedal is experienced, bleed all brakes.

s. **Bleeding Brakes.** Remove screws holding brake master cylinder inspection cover to toeboard between foot pedals and remove cover. Reach through hole, and clean around master cylinder filler cap. Remove cap and fill master cylinder with brake fluid. Replace cap temporarily. Clean all bleeder connections at wheel cylinders (fig, 90). Attach bleeder hose to *right rear* wheel cylinder bleeder screw, and place end in a glass jar or bottle so that the end is submerged in brake fluid. Open bleeder screw a three-quarter turn. Press brake pedal by hand, allowing it to return slowly. Continue action until air bubbles cease to appear at end of bleeder hose. Tighten bleeder screw and remove hose. Follow the same procedure on the *right front* brake, then the *left rear,* and finally, the *left front* brake. Replenish master cylinder brake fluid supply. Install filler cap and inspection cover.

152. PARKING (HAND) BRAKE.

a. Adjustment. Place hand brake grip in released position. Check brake levers to see that cable is free and released. Remove lock wire from anchor adjusting screw (fig. 91). Place 0.005-inch thickness gage between band and drum at anchor screw, and adjust screw to secure clearance. Install lock wire. Tighten adjusting nut until brake band is tight around drum. Loosen bracket bolt lock nut and rear nut. Back nut off two turns and set lock nut. Loosen adjusting nut so that brake band has approximately 0.010-inch clearance on drum.

b. Removal of Parking (Hand) Brake Band. Remove anchor bolt lock wire. Remove anchor bolt/Remove bracket bolt. Remove cotter pin from brake cam clevis pin and remove pin. Remove brake band adjusting nut. Remove brake adjusting bolt and spring. Remove retracting spring, and remove brake band assembly.

e. Installation of Parking (Hand) Brake Band. Install brake band on drum. Install brake release spring and adjusting bolt. Install clevis pin in brake cam and head of adjusting bolt. Install cotter pin in clevis pin. Install brake band adjusting nut. Install bracket bolt and nuts. Install anchor bolt. Adjust brake (subpar. a above). Install retracting spring.

Section XXVIII

SPRINGS AND SHOCK ABSORBERS

153. DESCRIPTION AND DATA.

a. **Description.** The springs (figs. 92 and 93) are of the semi-elliptic type with the second leaf wrapped around the spring eye of the first (main) leaf. The front springs appear to be identical, but

RA PD 305265

Figure 92—Left Front Spring

have different load carrying ability. The left spring has an "L" painted on the underside of the second leaf at the front end. Four spring leaf clips keep the leaves in alinement, and hold the leaves together to take the rebound. The front spring is shackled at the front end; the rear spring is shackled at the rear end. A spring pivot bolt attaches the opposite end of the spring to the frame. The shackles are of the threaded U-bolt type with threaded bushings having right- and left-hand threads. The left-hand threaded shackle ends and bushings are used in the spring eye of the left front spring and right rear spring. Left-hand threaded shackles have a small forged boss on the lower shank of the shackle. The left-hand threaded bushings have a groove cut around the hexagon head. The left front spring is equipped with a torque reaction spring to stabilize the front axle in extremely rough service. The shock absorbers are of the hydraulic-cylinder type, direct-acting, two-way control, adjustable, and refill-able.

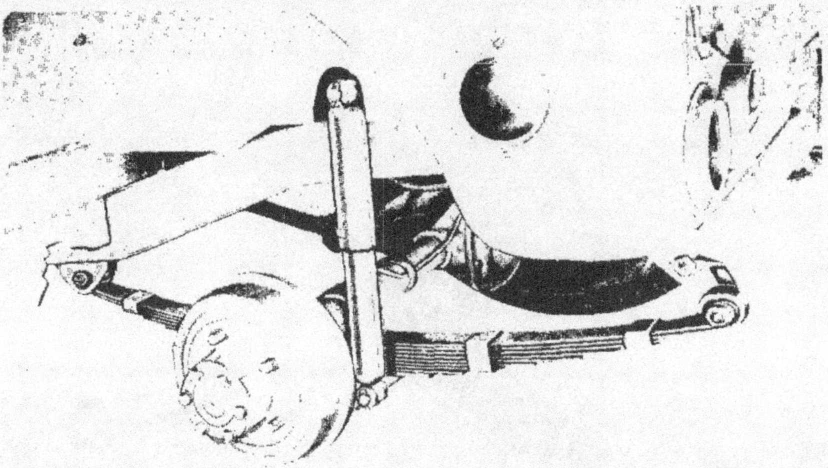

RA PD 305266

Figure 93—Left Rear Spring

b. **Data.**

Front Springs

Length–center line of eyebolts	36¼ in.
Width	1¾ in.
Number of leaves	8
Spring center bolt	At center
Spring eye bushed	Rear

Rear Springs

Length–center line of eyes	42 in.
Width	1¾ in.
Number of leaves	9
Spring center bolt	At center
Spring eye bushed	Front

Shock Absorbers	
Type	Hydraulic
Action	Double
Length—compressed—front	10 9/16 in.
Length—compressed—rear	11 9/16 in.
Length—extended—front	16 1/8 in.
Length—extended—rear	18 1/8 in.
Adjustable	Yes
Refillable	Yes
Mountings	Rubber bushings

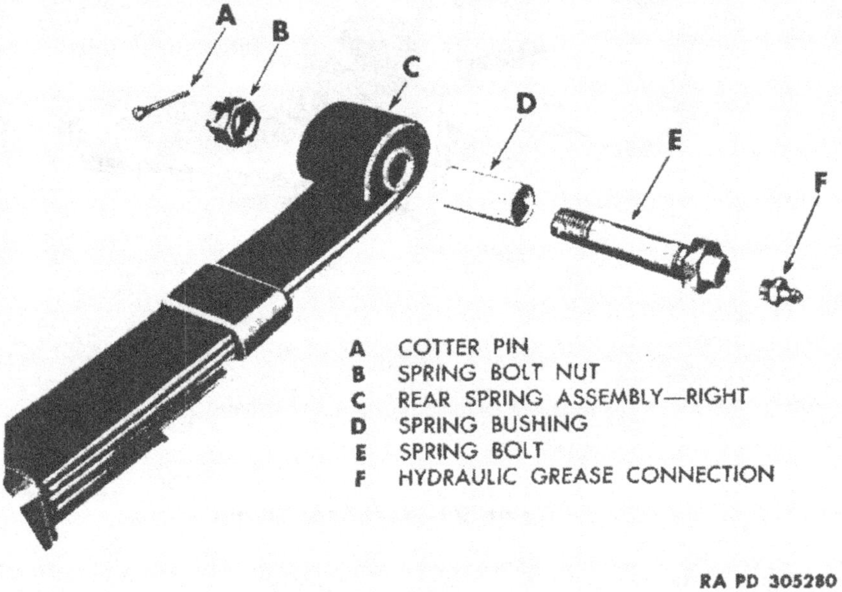

Figure 94—Right Rear Spring Bolt

154. MAINTENANCE.

a. The springs and shock absorbers should be inspected periodically in accordance with preventive maintenance (par. 23). Lubricate springs to prevent breakage, and excessive wear of spring pivot bolts and shackles (par. 18). Spring bushings and shackles must be free to move. Adjust shock absorbers correctly (par. 157). Replace worn or damaged shock absorber mounting bushings (par. 157).

155. SPRING SHACKLES AND BOLTS.

a. **Removal of Spring Bolt.** Raise vehicle frame until tires just rest on floor. Pull cotter pin in spring pivot bolt nut. Remove nut and drive out bolt (fig. 94).

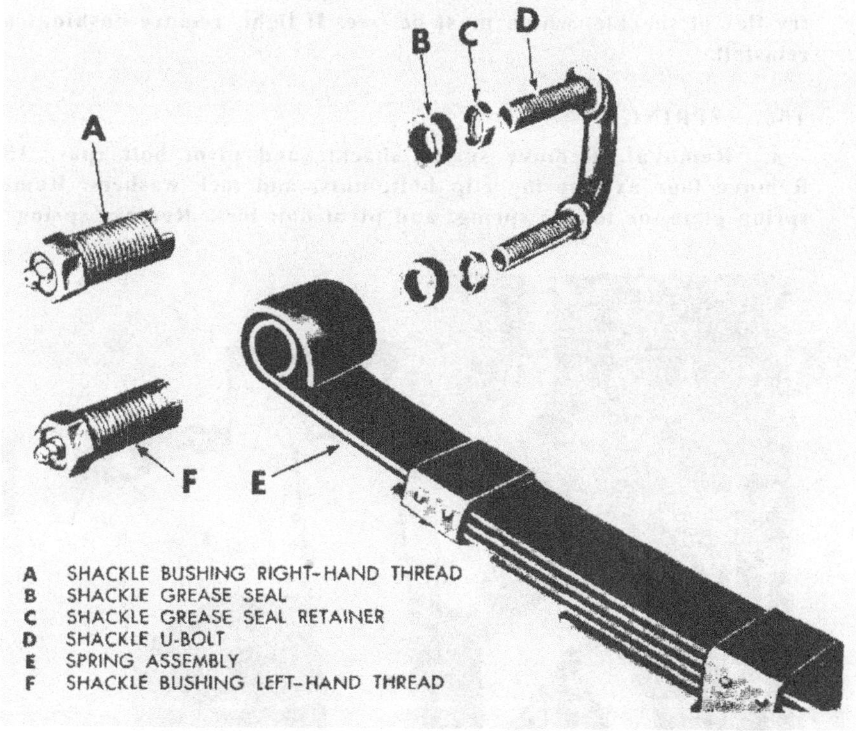

Figure 95, *Left Front Spring Shackle*

RA PD 305279

Figure 95—Left Front Spring Shackle

b. **Installation of Spring Bolt.** Line up holes in spring bracket and spring. Drive spring pivot bolt into place with oil groove up. Install nut and cotter pin. Lubricate with high pressure grease gun. Lower vehicle to floor.

c. **Removal of Spring Shackle.** Raise vehicle frame until tires just rest on floor. Remove shackle bushings (fig. 95). NOTE: *Left-hand threaded bushings are used in spring end of shackles on left front spring and right rear spring.*

d. **Installation of Spring Shackle.** Install shackle grease seal and retainer over threaded end, and up to the shoulder. Insert shackle through frame bracket and eye of spring, giving due attention to right- and left-hand threads. Hold shackle tightly against frame, and start upper bushing on shackle. Run in about half-way, then start lower bushing, holding shackle tightly against spring eye. Run bushing in about half-way. Then alternately tighten bushings until upper bushing is tight against frame bracket, and lower bushing hexagon head is about 1/32 inch away from spring eye. Lubricate bushings, and try flex of shackle, which must be free. If tight, remove bushings and reinstall.

156. SPRINGS.

a. Removal. Remove spring shackle and pivot bolt (par. 155). Remove four axle spring clip bolt, nuts, and lock washers. Remove spring plate, or torque spring, and pivot bolt lock. Remove spring.

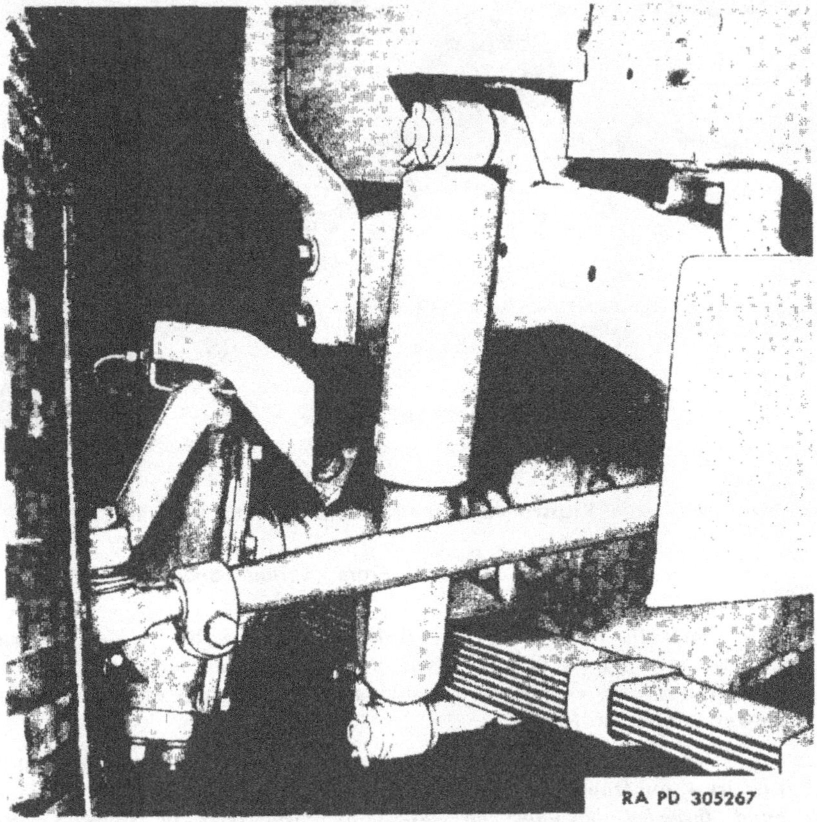

Figure 96—Shock *Absorber*

b. Installation. Install spring pivot bolt (par. 155 b). Install shackle (par. 155 d). Raise vehicle, and place center bolt in spring saddle on axle. Install axle spring clips and nuts. NOTE: *Axle spring clip nut torque wrench reading should be 50 to 55 foot-pounds; spring pivot bolt nut, 27 to 30 foot-pounds; torque reaction spring bolt, 60 to 65 foot-pounds.*

157. SHOCK ABSORBERS.

a. Removal. Pull cotter pins holding upper and lower washers against rubber bushing on mounting brackets. Remove washers, and pull off shock absorbers and rubber bushings (fig. 96).

b. Installation. Check shock absorber adjustment; compress shock absorber, and turn ore end to engage adjusting keys in slots. Turn end in clockwise direction until limit of adjustment is reached, then turn end counterclockwise two turns for average adjustment. NOTE: *Turn end clockwise for firmer control, and counterclockwise for softer control, allowing faster spring action.* Install inner mounting rubber bushing on upper and lower bracket pins. Install shock absorber. Install outer bushing and flat washer. Use bushing compressor (41-C-2554-400) to compress bushing, and install cotter pin. Spread both ends of cotter pin to hold washer evenly in proper position.

Section XXIX

STEERING GEAR

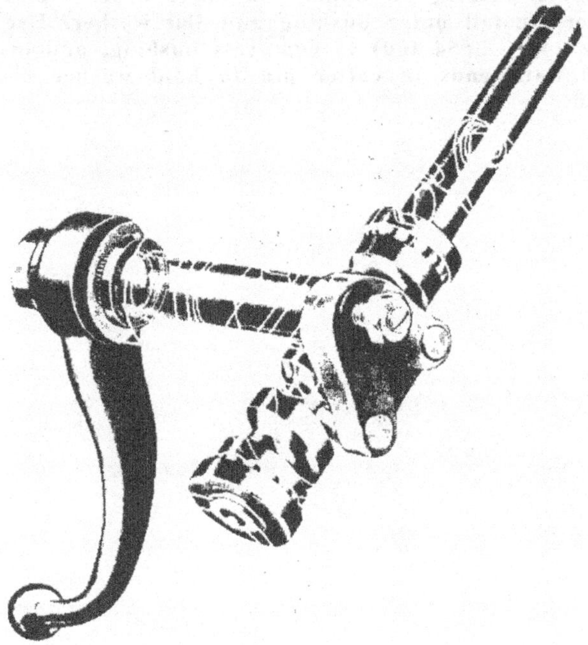

RA PD 305269

Figure 97—Steering Gear—Phantom View

158. DESCRIPTION AND DATA.

a. Description. The steering gear (figs. 97 and 99) is of the conventional type, mounted on the left frame side member, and connected to the front axle steering ball crank by a Pitman arm and steering connecting rod (fig. 98). The steering gear is of the cam and lever type with a variable-ratio cam. The steering wheel is of the

3-spoke, safety type, with 17 1/4-inch diameter. The steering connecting rod is of the adjustable, ball-and-socket type.

b. Data.

Make and model . . . . Ross T-12
Type . . . . Cam and twin pin lever
Ratio . . . . Variable; 14-12-14 to 1
Wheel . . . . 3-spoke; safety type; 17 1/4 in.

159. MAINTENANCE.

a. Maintenance consists primarily of proper lubrication (par. 18) and periodic inspection in accordance with preventive maintenance

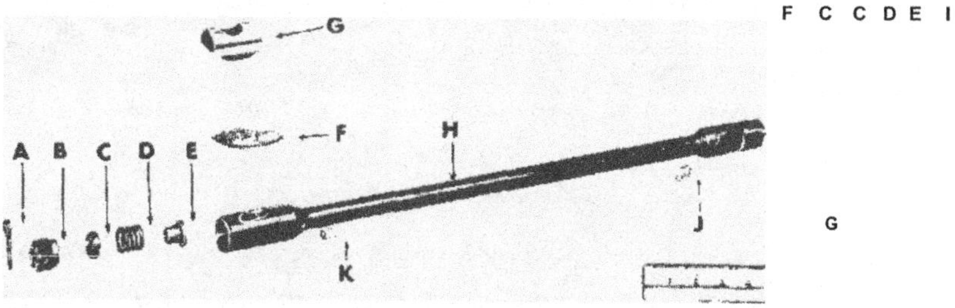

A COTTER PIN
B ADJUSTING PLUG-LARGE
C BALL SEAT
D SPRING
E SAFETY PLUG
F DUST COVER
G DUST COVER SHIELD
H CONNECTING ROD ASSEMBLY
I ADJUSTING PLUG-SMALL
J HYDRAULIC GREASE FITTING
K HYDRAULIC GREASE FITTING

RA PD 305278

Figure 98—Steering Connecting Rod, *Disassembled*

procedures (par. 23) to include the Pitman arm and steering connecting rod. A systematic inspection for steering troubles is as follows:

(1) Equalize tire pressures and set car on level floor.

(2) Inspect king pin and wheel bearing for looseness.

(3) Check wheel run-out.

(4) Check for spring sag.

(5) Inspect brakes and shock absorbers.

(6) Check steering assembly and connecting rod.

(7) Check toe-in.

(8) Check toe-out on turns.

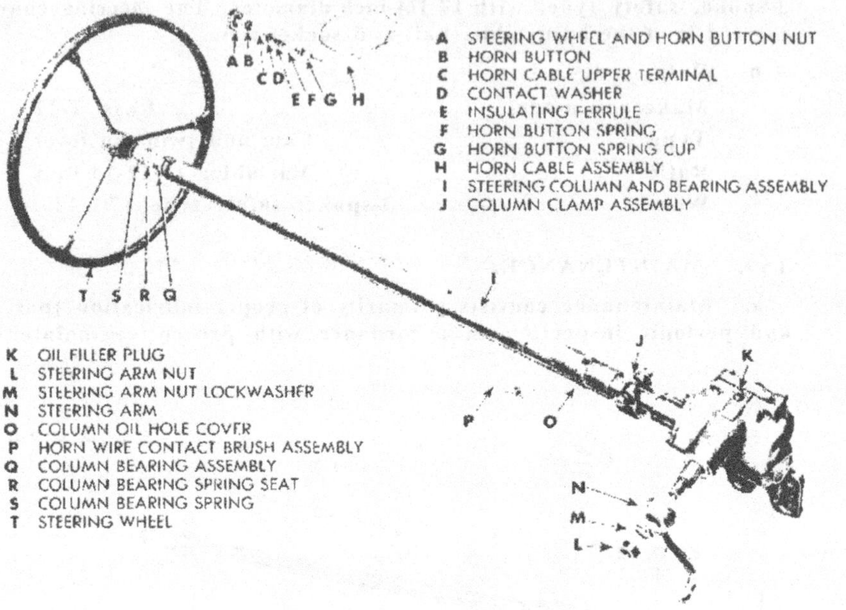

Figure 99—Steering Gear, Disassembled

(9) Check tracking of front and rear axle.

(10) Check frame aiinement.

b. If steering difficulty is experienced after checking and correcting the above items, report to higher authority, because the trouble may be due to wheel balance, caster, camber, or king pin inclination.

160. STEERING CONNECTING ROD.

a. Removal. Pull cotter pin at each end of rod. Unscrew plugs and remove rod.

b. Installation. Correct end of connecting rod to be attached to front axle bell crank will have the lubrication hydraulic fitting to the right. Install safety plug, spring, and ball seat in this end of rod. Install rod on bell crank ball. Install adjusting plug. Screw plug in firmly against ball, back off one-half turn, and lock with cotter pin. Insert ball seat in other end of rod. Install rod on steering Pitman arm. Install second ball seat, spring, safety plug, and adjusting plug in order. Screw plug in firmly against ball, back off one-half turn, and lock with cotter pin. Lubricate with high pressure gun.

161. STEERING WHEEL.

a. Removal. Raise hood, remove horn wire at steering post terminal, and tape end so it will not ground. Remove steering wheel nut, and lift off horn button. Pull steering wheel off with steering wheel puller.

b. Installation. Set front wheels straight ahead. Install steering wheel so that one spoke of wheel is in vertical position above steering post. Drive wheel down on post. Install horn button and steering wheel nut. Untape horn wire, attach to steering post terminal, try horn, and lower hood.

162. STEERING PITMAN ARM.

a. Removal. Remove Pitman arm nut and lock washer. Remove Pitman arm by using wedge type Pitman arm remover, or as follows: Drive a chisel between the arm and the steering gear case at the front side, and using a bar, strike rear side of arm to loosen it on the tapered serrations. Pull cotter pin in rear end of steering connecting rod, and remove adjusting plug. Take steering connecting rod off Pitman arm ball.

b. Installation. Turn steering wheel maximum distance to right; turn wheel to left, and count turns. Turn wheel to right exactly one-half of the turns. Install steering connecting rod on Pitman arm (par. 160 b). Set front wheels in straight-ahead position, and install Pitman arm on steering gear. Install lock washer and nut. Tighten nut securely. Lubricate connecting rod hydraulic fitting.

163. STEERING GEAR.

a. Removal. Raise hood and tie to windshield. Remove battery ground cable at post on battery. Release headlight bracket wing nut and tilt headlight away from fender. Remove headlight wires from junction block on fender splasher. Remove blackout headlight wire clip on fender. Remove horn from bracket. Remove headlight wires from junction block on dash. Disconnect blackout driving light wire from slip connector at dash. Remove two screws attaching horn wire contact brush assembly to steering column. Loosen steering column clamp bolt. Remove bolts attaching fender to body, frame, and radiator grille. Remove fender support bolts in frame and remove fender. Remove cotter pin from rear end of steering connecting rod, unscrew adjusting plug, and remove from Pitman arm ball. Remove steering wheel nut and horn button. Pull steering wheel with steering wheel puller. Remove steering column bracket. Remove bolts in steering column floor seal, and remove seal and retainer. Pull steering column up off tube. Remove steering column to frame bolts. Lower upper end of steering column, and lift lower end out over frame side member.

b. Installation. Check steering gear lubricant; replenish if necessary. Insert upper end of steering gear through toeboard, and position in chassis. Install steering gear to frame bolts, but do not tighten. Install steering column floor seal, retainer, and screws. Install steering column over tube, with horn contact brush onening up. Tighten steering column clamp. Install horn wire contact brush, and tighten screws. Attach steering column bracket, and tighten steering gear to frame bolts. Install steering wheel (par. 161 **b**). Install steering connecting rod (par. 160 **b**). Install fender and support bolts, also screws and bolts to frame, body, and radiator grille. Install blackout headlight wire clip on fender. Attach headlight wires to junction block on fender splasher. Connect blackout headlight wire at slip connector. Attach headlight wire to junction block on dash. Connect blackout driving light wire in slip connector at dash. Install horn on bracket. Tilt headlight, and tighten wing nut. Attach battery cable. Check operation of horn and lights. Lower hood and lock.

o. Adjustment. Loosen lock nut on side adjusting screw. With front wheels straight ahead, adjust screw for minimum backlash of studs in cam groove. Tighten lock nut. For other adjustments, report to higher authority.

Section XXX

BODY AND FRAME

164. DESCRIPTION AND DATA.

a. Description. The body (figs. 1 to 4) is of the open type, identified by a name plate located on the instrument panel (figs. 5 and 6). There are two individual tubular frame front seats and a rear seat. The left front seat cushion can be raised to fill the fuel tank; the right front seat can be raised forward for stowage of the vehicle removable top, curtains, and windshield and light covers. The rear seat can be raised to reach the tire pump. Tools and accessories are stowed in two compartments in the rear corners of the body. The windshield is equipped with dual, hand-operated wipers, and can be opened forward or folded down on top of the hood. A fire extinguisher and an adjustable rear vision mirror are mounted on the left side of the cowl. Safety straps are provided in the entrance ways. A rifle holder is mounted on the lower panel of the windshield over the instrument panel. A strap and sheath carry a shovel and ax on the left side of the body. Hand grips on the side of the body facilitate lifting. The fuel tank sets in a sump in the floor pan under the driver's seat. The vehicle top is supported by top bows which can be folded down along the body sides to form a hand rail. A fuel can rack, trailer connection, and spare tire and wheel are mounted on the body rear panel. The chassis has five cross members; the rear intermediate cross member having a gun platform. Box-type, reinforced frame side members are used for maximum strength. Bumpers at the front and rear, and a radiator guard provide protection against damage. A pintle hook at the rear provides a means of hauling a trailed load.

b. Data.

Body type	Open	Windshield type	Folding
Driver's position	Left side	Cross members	5
Chassis frame type	Double drop		

165. MAINTENANCE.

a. General maintenance of the body requires periodic tightening of all loose parts, and lubrication of wearing parts. Keep the body clean and touch up bare spots to prevent rust. Keep the sump under the fuel tank free of dirt, stones, and water. Keep the sump front drain hole cover on so that dirt and water thrown by the front wheel will not enter. Keep the rear drain hole cover in the glove compartment in the instrument panel except when crossing water, when it must be installed. Water in the body can be drained by removing drain plugs in the floor at the side of the cowl. Chassis maintenance concerns primarily, proper lubrication of connecting parts (par. 18).

166. INSTRUMENTS.

a. Procedure for the removal and installation of the various panel instruments is identical, and as follows:

b. Removal. Remove battery ground cable at battery post as a safety precaution. Remove connecting wires or tubes. Remove two nuts holding retaining clamp in place, and remove instrument through face of instrument panel.

c. Installation. Install instrument in place in panel. Install retaining clamp and nuts. Attach tubes or wires.

167. SEATS AND CUSHIONS.

a. Removal of Seat Cushions and Backs. Remove five screws holding front seat cushion to frame at rear side, and remove cushion. Remove 10 screws holding seat back to frame, and remove seat back. Lift up back edge of rear seat cushion, remove five screws holding front edge of cushion to frame, and remove cushion. Remove five screws in top edge of seat back and two in lower edge, and remove seat back.

b. Installation of Seat Cushions and Backs. Place rear seat back in position, and install two lower screws. Pull edge of seat back up in place, and install five screws in top side. Place rear seat cushion in position, top side down. Install screws, and turn cushion over into place. Place front seat back in position, and install screws. Place seat cushions in position, and install screws.

c. Removal of Front Seats. Remove three screws holding back of driver's seat to floor. Remove screw in wheel housing holding seat. Remove two bolts holding front of seat frame to floor, and lift out seat. Remove two bolts holding right front seat bracket to floor, and lift out seat.

d. Installation of Front Seats. Place seat in position. Install bolts in place at front of seat. On driver's seat install screws and bolts holding seat back to floor and wheel housing.

e. Removal of Rear Seat. Pull up front edge of seat to fold seat. Remove bolt in tool compartment holding retainer bracket at seat bracket. On same side remove two bolts holding seat back bracket. Raise end of seat and lift out.

f. Installation of Rear Seat. Place seat in position in brackets. Install retainer bracket. Install seat back bracket and bolts.

168. WINDSHIELD WIPER.

a. Removal. Remove nuts holding wiper handle, and remove handles. Remove plain washer, and remove wiper blades.

b. Installation. Install blades and arms in place through windshield frame. Install plain washers, handles, and nuts.

169. WINDSHIELD.

a. Removal. Unhook windshield clamps on instrument panel (fig. 5). Remove wing screws at sides of cowl, and lift off windshield.

b. Installation. Place windshield in position, and install wing screws at sides of cowl. Clamp windshield to instrument panel.

170. TOP.

a. Installation. Loosen the two wing screws at the pivot brackets (fig. 4). Slide tubular bows back out of front bracket. Install front ends in rear brackets, and tighten winged screws. Allow front bow to drop down over seats. Remove top from under right front seat. Attach top to fasteners at top of windshield. Stretch top over bow and down to body back panel. Place straps in metal loops, and attach to body panel; stretch top, and buckle straps. Raise front bow into position at bow flaps, and snap flaps around bow. The curtains are attached in the conventional way with snap fasteners.

b. Removal. Remove curtains by releasing snap fasteners. Unsnap bow flaps, and lower front bow on front seat. Unbuckle top straps at body rear panel. Unsnap top at top of windshield, and remove. Fold top and stow under right front seat. Loosen wing screws in top rear brackets. Fold front bow against rear bow. Raise bows out of rear brackets, and insert lower ends in front brackets. Tighten rear bracket screws.

171. RIFLE HOLDER.

a. Removal. Swing the rifle bumper to the right, at the right end of holder, and remove rifle. Remove two bolts holding rifle holder to windshield lower panel, and remove holder (fig. 5).

b. Installation. Place rifle holder in position on windshield panel with butt end to the left, insert bolts, and tighten securely. Swing rifle bumper to the right. With barrel up, insert butt end of rifle in holder at the left. Push rifle up against spring pressure, and turn bumper to left under rifle.

172. SHOVEL AND AX.

a. **Removal.** Release straps and remove shovel or ax individually.

b. **Installation.** Turn bit, or blade, of ax up. Insert handle in front clamp. Insert blade in sheath. Pull up clamp under ax head, and strap in place. Turn face of shovel against cowl and place in strap on cowl side. Wrap fabric strap, through handle, over grip, between grip and side of body, through loop, over outside of grip, and buckle. NOTE: *This will hold the shovel forward in the strap on the cowl side (fig. 100).*

173. HOOD.

a. **Removal.** Unhook hood and raise against windshield. Remove screws in hinge at cowl, and disconnect bond strap.

RA PD 305271

Figure 100—Shovel *and Ax Mounting*

b. **Installation.** Place hood in position and install hinge screws in cowl, but do not tighten. NOTE: *Install bonded screws last as follows:* Install flat washer on screw. Install screw through bond strap. Install flat washer. Install washer between hinge and hood. Install screw through hinge, and tighten to cowl. Lower hood for alinement. Raise hood and tighten screws. Lower hood and hook down both sides.

174. RADIATOR GUARD.

a. **Removal.** Raise hood. Remove headlight hinge bolts. Remove headlight wire clips on guard. Remove blackout headlight wire clip on left front fender. Remove wires from slip connector at left front fender. Remove fender to guard bolts. Remove guard from chassis. Remove blackout headlight wires, clips, and loom. Remove rubber shield at headlight. Remove blackout headlight nut and washer, and remove both light and wire assemblies.

b. Installation. Install both blackout headlight, spacer, and wire assemblies on guard. Install washer and nut. Install shield over wire. Install loom on wire. Install wire clips on loom and clips on guard. Install frame bolts in guard. Install guard on chassis. Install fender to guard bolts loosely. Install guard to frame bolt nuts. Tighten fender to guard bolts. Install headlight hinge bolts. Install headlight wire clips on guard. Install blackout headlight wire clip on fender. Connect wires to slip connector. Check operation of lights. Lower hood and lock down.

175. FENDERS.

a. Removal of Right Front Fender. Raise hood. Loosen wing nut on headlight bracket, and tilt light up. Remove battery to front fender strap. Remove battery ground cable at battery post. Remove voltage regulator bolts. Remove fuel line clip to fender. Remove hood catch assembly. Remove fender to radiator guard bolts. Remove fender bolts in support, body, and frame. Remove fender.

b. Installation of Right Front Fender. Place fender on chassis. Install one fender to body bolt. Install one fender to guard bolt. Install other fender bolts and tighten all. Install fuel line clip. Install voltage regulator. Install battery to fender strap. Install battery cable. Position headlight, and tighten wing nut. Install hood catch. Lower hood and lock.

c. Removal of Left Front Fender. Raise hood. Remove headlight bracket wing nut, and tilt lamp up. Remove two wire clips on splasher. Remove wires from junction block, and slip connector at dash. Remove blackout driving light clip on top of fender. Remove three bolts in blackout driving light bracket. Remove blackout driving light wire grommet and clips from fender. Remove junction block on fender. Remove blackout headlight wire clip. Remove hood catch. Remove fender shield to frame bolt. Remove bolts between fender and guard support, body, and frame. Remove fender.

d. Installation of Left Front Fender. Place fender on chassis. Install guard upper bolt. Install all fender bolts loosely. Install blackout headlight wire clip (front) on fender. Install junction block to fender. Install blackout driving light wire through fender and splasher. Install three bolts in blackout driving light bracket and fender. Install blackout headlight wire clip on fender. Install wire grommet in fender. Install two wire clips to fender splasher. Connect wires to junction block at dash and slip connector. Install hood catch. Place headlight in position, and secure with wing nut. Check operation of lights. Lower hood and lock down.

Section XXXI

RADIO INTERFERENCE SUPPRESSION SYSTEM

176. DESCRIPTION.

a. Description. Radio noise suppression is the elimination, or minimizing, of electrical disturbances which interfere with radio reception, or disclose the location of the vehicle to sensitive electrical detectors. Electrical disturbances or radio frequency waves may originate as static discharges between adjoining parts of the vehicle, or may be given off by the electrical systems during operation of the vehicle. These waves are actually radiated as disturbing signals that interfere with any radio receiving apparatus that may be operating in the vehicle or immediate vicinity. Each disturbance (at plugs, breaker points, generator brushes, or elsewhere) creates a surge of electricity, which produces interfering radio waves. Their origin can generally be determined by the nature of the noise heard in the receiver. Radio interference suppression, therefore, involves the suppression of these waves at their sources, or confining them within an area where they cannot be picked up by the antenna of a radio-equipped vehicle. Suppression is accomplished by the use of resistor-suppressors, and condensers. In addition, the hood and other metal parts in the vicinity of the engine are made to form a shield by the use of internal-external toothed lock washers and bond straps; thus, the hood and side panels form a box within which radio frequency waves are confined to prevent their acting on the antenna of receiving equipment. Wiring that may carry interfering surges to a point where interference will affect radio reception, is shielded. In attaching condensers and bond straps, the lock washers must be placed between the parts to be grounded, and tinned spots must be cleaned, but not painted. This is necessary to obtain good connections between the component parts, and to permit electrical energy to dissipate without causing electrical disturbances. The suppression components have no effect on engine performance as long as they are maintained in good condition. The sources of electrical noise interference may be basically divided into three groups: the ignition system, including coil, distributor, and spark plugs; the generator system, including generator and regulator; and the wiring.

177. DATA.

a. Ignition (both high-tension and primary-circuit suppression).

(1) High-tension suppression is of the resistor-suppressor type and consists of:

(a) Coil to distributor high-tension wire at distributor, resistance 10,000 ohms.

(b) Spark plug high-tension wire at spark plugs, resistance 10,000 ohms.

(2) Primary circuit suppression is of the capacitive-filter type and consists of:

(a) Ignition coil terminal (+) to cylinder block, capacity 0.10 microfarad.

(b) Ignition switch terminal (lower) to instrument panel, capacity 0.01 microfarad.

1». Charging Circuits (includes generator, regulator, ammeter, and battery).

(1) Generator suppression is of the capacitive-filter type and consists of generator armature terminal (A) to ground on generator, capacity 0.10 microfarad.

(2) Regulator suppression is of the capacitive-filter type and consists of:

(a) Regulator field terminal (F) to ground, capacity 0.01 microfarad.

(b) Regulator field terminal (B) to ground, capacity 0.25 microfarad.

c. Miscellaneous Circuits (including radio box and starting circuits).

(1) Radio terminal box suppression is of the capacitive-filter type and consists of:

(a) Radio box terminal to ground, capacity 0.50 microfarad.

(b) Starting switch battery terminal to floor, capacity 0.50 microfarad.

d. Bonding.

(1) BOND STRAPS (ground straps) (figs. 101 and 102). Bond straps are installed from:

(a) Hood to dash, right side (D).

(b) Hood to dash, left side (I).

(c) Cylinder head stud to dash (H).

(d) Cables (hand brake, speedometer, heat indicator) to dash (E).

(e) Generator mounting bolt to cranking motor bracket to engine support bracket (A).

(f) Generator to regulator wire shield to ground on generator and regulator (B).

(g) Front engine bracket to frame, left side (J).

(h) Hood ground to grille, left side (L).

(i) Hood ground to grille, right side (N).

(j) Radio terminal box to ground wire (F).

(2) TOOTHED LOCK WASHERS (figs. 101 and 102). Toothed lock washers are supplied from:

(a) Radiator to frame, right side (O).

(b) Radiator to frame, left side (K).

(c) Body bracket ground to frame, right side (S).

(d) Body bracket ground to frame, left side (R).

(e) Fender splasher ground to frame, right side (U).

(f) Fender splasher ground to frame, left side (T).

(g) Air cleaner mounting (C).

(h) Body hold-down bolts (G).

(i) Radiator grille to cross member (M).

(j) Fender to cowl, left side (P).

(k) Fender to cowl, right side (Q).

178. TESTS.

a. General. Electrical disturbances which cause radio interference are loose bonds, loose toothed lock washers, broken or cracked resistor-suppressors, loose connections, or faulty filters. Following are tests which can be made to determine the cause of interference. The radio equipment in the vehicle may be used as a test instrument to localize troubles, and to determine when faulty parts or conditions have been eliminated or corrected. If the vehicle has no radio equipment, utilize a radio-equipped vehicle placed about 10 feet from the vehicle under test. Here the cooperation of the radio operator is required. Determine the circuits causing the noise by checking as follows:

(1) Operate engine while listening to radio. A regular clicking which varies with engine speed, and ceases the instant the ignition is shut off, is caused by the ignition circuit.

(2) An irregular clicking which continues a few seconds after the ignition is shut off, is caused by the regulator.

(3) A whining noise which varies with engine speed, and continues a few seconds after the ignition is shut off, is caused by the generator.

b. Noise Caused by Ignition Circuit.

(1) Make certain ignition system is functioning properly (section XV). Improper plug gaps, late timing, poor adjustment of breaker points, and damaged or worn distributor, will affect the suppression system.

(2) Inspect resistor-suppressors in spark plug leads. Replace any that are scorched, cracked, or otherwise faulty. Be sure wires are screwed in tightly.

(3) Inspect resistor-suppressor at distributor. Replace if necessary.

(4) Inspect and tighten all bonds in engine compartment.

(5) Inspect capacitive-type filters in primary circuit at ignition coil and ignition switch. Make certain mounting bolts are tight. Replace filter and test for noise.

c. Noise Caused by Regulator.

(1) Check all connections to regulator.

(2) Check capacitive-type filter mounting bolts for tightness and correct placement of lock washers.

(3) Check regulator mounting bolts for tightness and correct placement of lock washers.

(4) Test for noise.

(5) If noise is still present, replace battery circuit filter attached to regulator (B) terminal. Test for noise.

(6) Replace field circuit filter attached to regulator (F) terminal. Test for noise.

(7) Replace armature circuit filter attached to generator (A) terminal.

(8) Test for noise.

d. Noise Caused by Generator.

(1) Check to make certain there is no excessive sparking at brushes. Correct if necessary.

(2) Inspect filter mounting, and check placement of lock washers. Tighten.

(3) Inspect ground strap.

(4) Replace filter.

(5) Test for noise.

e. Noise Caused by Miscellaneous Circuits (radio box and starting switch).

(1) Inspect mounting of filter attached to circuit. Tighten and test.

(2) Replace filter.

(3) Test for noise.

f. Noise Observed While Vehicle Is in Motion, but Not When Stopped.

(1) Inspect and tighten all body bonds (par. 177 d (1) above and figs. 101 and 102).

(2) Inspect and tighten all points where toothed lock washers are used (par. 177 d (2) above and figs. 101 and 102).

(3) Test for noise.

179. MAINTENANCE.

a. General. General maintenance of the radio suppression system (par. 23 a (5), item 104) must be made in connection with preventive maintenance items, particularly in regard to spark plugs, distributor and wires, late ignition timing, generator brushes, loose switch contacts, and discharged battery causing high generator charging rate.

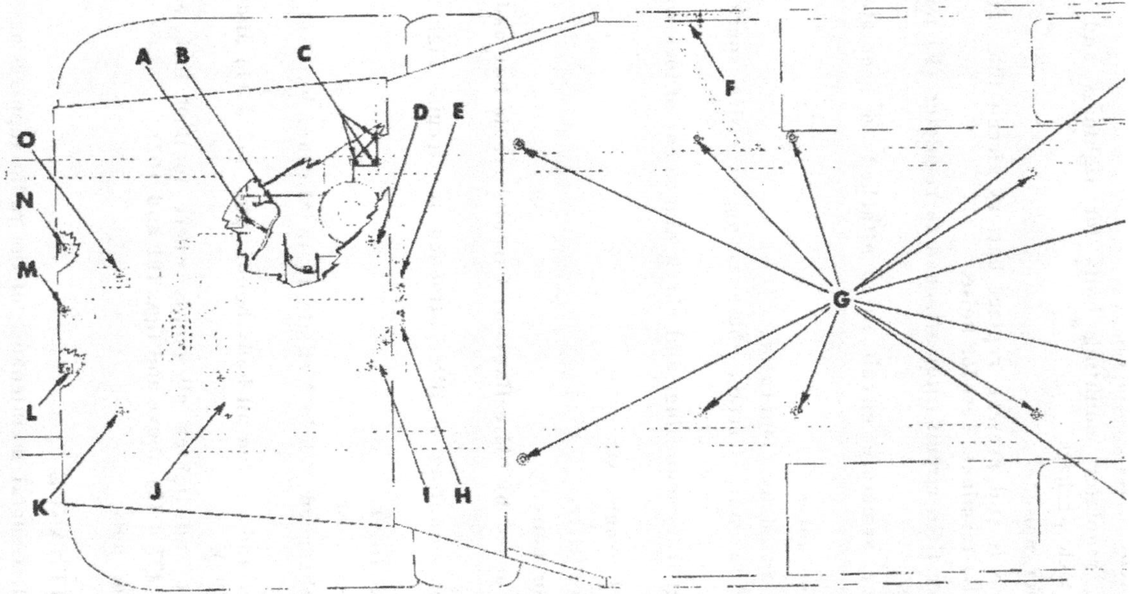

A—GENERATOR MOUNTING BOLT TO CRANKING MOTOR BRACKET TO ENGINE SUPPORT BRACKET
B—GENERATOR TO REGULATOR WIRE SHIELD TO GROUND ON GENERATOR AND REGULATOR
C—AIR CLEANER MOUNTING
D—HOOD TO DASH—RIGHT SIDE
E—CABLES (HAND BRAKE, SPEEDOMETER, HEAT INDICATOR), TO DASH
F—RADIO TERMINAL BOX TO GROUND WIRE
G—BODY HOLD-DOWN BOLTS
H—CYLINDER HEAD STUD TO DASH
I—HOOD TO DASH—LEFT SIDE
J—FRONT ENGINE BRACKET TO FRAME—LEFT SIDE
K—RADIATOR TO FRAME—LEFT SIDE
L—HOOD GROUND TO GRILLE—LEFT SIDE
M—RADIATOR GRILLE TO CROSSMEMBER
N—HOOD GROUND TO GRILLE—RIGHT SIDE
O—RADIATOR TO FRAME—RIGHT SIDE

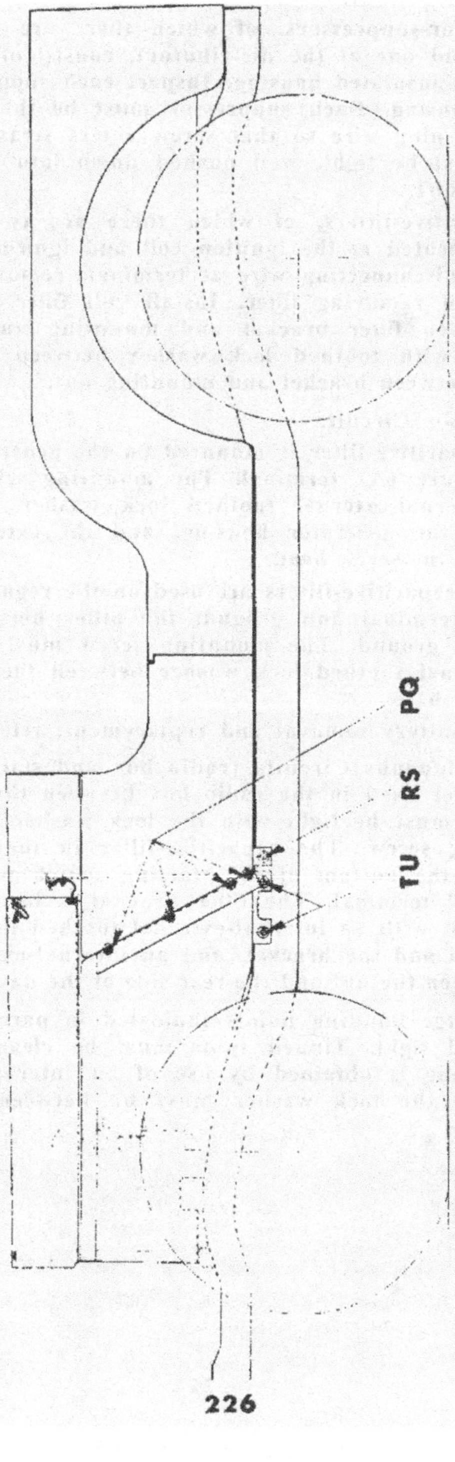

P—FENDER TO COWL—LEFT SIDE
Q—FENDER TO COWL—RIGHT SIDE
R—BODY BRACKET GROUND TO FRAME—LEFT SIDE
S—BODY BRACKET GROUND TO FRAME—RIGHT SIDE
T—FENDER SPLASHER GROUND TO FRAME—LEFT SIDE
U—FENDER SPLASHER GROUND TO FRAME—RIGHT SIDE

RA PD 334758

Figure 102—Location of Bond Straps and Fastenings (2)

b. Ignition Circuits.

(1) Resistor-suppressors, of which there are five (one at each spark plug and one at the distributor), consist of a high resistance element in an insulated housing. Inspect each suppressor for cracked or broken housing. Each suppressor must be threaded tightly into end of spark plug wire so that screw enters strands of cable. Wire terminals must be tight, well pushed down into place, and free of corrosion or dirt.

(2) Capacitive-filters, of which there are two in the ignition circuit, are located at the ignition coil and ignition switch. Replace if faulty by disconnecting wire at terminal, removing mounting nut or screw, and removing filter. Install coil filter with toothed lock washer between filter bracket and mounting nut. Install ignition switch filter with toothed lock washer between filter bracket and panel; also between bracket and mounting nut.

c. Charging Circuit.

(1) A capacitive-filter is mounted on the generator, and attached to the armature (A) terminal. The mounting screw must be tight with the internal-external toothed lock washer between the filter bracket and the generator housing, and the external toothed lock washer under the screw head.

(2) Two capacitive-filters are used on the regulator; one between the battery terminal and ground, the other between the field coil terminal and ground. The mounting screw must be tight with the internal-external toothed lock washer between the filter bracket and the regulator base.

(3) For battery removal and replacement, refer to paragraph 97.

d. Miscellaneous Circuits (radio box and starting circuits). The capacitive-filter used in the radio box between the terminal and the ground wire must be tight with the lock washer under the head of the mounting screw. The capacitive-filter in the starting circuit is mounted on the bottom of the starting switch with the connection on the "live" terminal. The filter bracket is located under the left mounting bolt with an internal-external toothed lock washer between the bolt head and the bracket, and an internal-external toothed lock washer between the nut and the rear side of the dash.

e. Bonding. Bonding points indicated in paragraph 117 d must be clean and tight. Tinned spots must be clean but not painted. Where bonding is obtained by use of an internal-external toothed lock washer, the lock washer must be between the parts to be grounded.

Section XXXII

SHIPMENT AND TEMPORARY STORAGE

180. GENERAL INSTRUCTIONS.

a. Preparation for domestic shipment of the vehicle is the same as preparation for temporary storage or bivouac. Preparation for shipment by rail includes instructions for loading and unloading the vehicle, blocking necessary to secure the vehicle on freight cars, number of vehicles per freight car, clearance, weight, and other information necessary to properly prepare the vehicle for rail shipment. For more detailed information, and for preparation for indefinite storage refer to AR 850-18.

181. PREPARATION FOR TEMPORARY STORAGE OR DOMESTIC SHIPMENT.

a. Vehicles to be prepared for temporary storage or domestic shipment are those ready for immediate service but not used for less than 30 days. If vehicles are to be indefinitely stored after shipment by rail, they will be prepared for such storage at their destination.

b. If the vehicles are to be temporarily stored or bivouacked, take the following precautions:

(1) LUBRICATION. Lubricate the vehicle completely (par. 18).

(2) COOLING SYSTEM. If freezing temperature may normally be expected during the limited storage or shipment period, test the coolant with a hydrometer, and add the proper quantity of anti-freeze compound to afford. protection from freezing at the lowest temperature anticipated during the storage or shipping period. Completely inspect the cooling system for leaks.

(3) BATTERY. Check battery and terminals for corrosion and if necessary, clean and thoroughly service battery (par. 97).

(4) TIRES. Clean, inspect, and properly inflate all tires. Replace with serviceable tires, tires requiring retreading or repairing. Do not store vehicles on floors, cinders, or other surfaces which are soaked with oil or grease. Wash off immediately any oil, grease, gasoline, or kerosene which comes in contact with the tires under any circumstances.

(5) ROAD TEST. The preparation for limited storage will include a road test of at least 5 miles, after the battery, cooling system, lubrication, and tire service, to check on general condition of the vehicle. Correct any defects noted in the vehicle operation, before the vehicle

is stored, or note on a tag attached to the steering wheel, stating the repairs needed or describing the condition present. A written report of these items will then be made to the officer in charge.

(6) FUEL IN TANKS. It is not necessary to remove the fuel from the tanks for shipment within the United States, nor to label the tanks under Interstate Commerce Commission Regulations. Leave fuel in the tanks except when storing in locations where fire ordinances or other local regulations require removal of all gasoline before storage.

(7) EXTERIOR OF VEHICLES. Remove rust appearing on any part of the vehicle with flint paper. Repaint painted surfaces whenever necessary to protect wood or metal. Coat exposed polished metal surfaces susceptible to rust, such as winch cables, chains, and in the case of track-laying vehicles, metal tracks, with medium grade rust-preventive lubricating oil. Close firmly all cab doors, windows, and windshields. Vehicles equipped with open-type cabs with collapsible tops will have the tops raised, all curtains in place, and the windshield closed. Make sure paulins and window curtains are in place and firmly secured. Leave rubber mats, such as floor mats, where provided, in an unrolled position on the floor; not rolled or curled up. Equipment such as Pioneer and truck tools, tire chains, and fire extinguishers will remain in place in the vehicle.

(8) **INSPECTION**. Make a systematic inspection just before shipment or temporary storage to insure all above steps have been covered, and that the vehicle is ready for operation on call. Make a list of all missing or damaged items, and attach it to the steering wheel. Refer to "Before-operation Service" (par. 13).

(9) ENGINE. To prepare the engine for storage, remove the air cleaner from the carburetor. Start the engine, and set the throttle to run the engine at a fast idle. Pour 1 pint of medium grade, preservative lubricating oil, Ordnance Department Specification AXS-674, of the latest issue in effect, into the carburetor throat, being careful not to choke the engine. Turn off the ignition switch as quickly as possible after the oi2 has been poured into the carburetor. With the engine switch off, open the throttle wide, and turn the engine five complete revolutions by means of the cranking motor. If the engine cannot be turned by the cranking motor with the switch off, turn it by hand, or disconnect the high-tension lead and ground it before turning the engine by means of the cranking motor. Then reinstall the air cleaner.

(10) BRAKES. Release brakes and chock the wheels.

c. Inspections in Limited Storage. Vehicles in limited storage will be inspected weekly for conditions of tires and battery. If water is added when freezing weather is anticipated, recharge the battery with a portable charger, or remove the battery for charging. Do not attempt to charge the battery by running the engine.

182. LOADING AND BLOCKING FOR RAIL SHIPMENT.

a. Preparation. In addition to the preparation described in paragraph **181,** when ordnance vehicles are prepared for domestic shipment, the following preparations and precautions will be taken:

(1) EXTERIOR, Cover the body of the vehicle with a canvas cover supplied as an accessory.

(2) TIRES. Inflate pneumatic tires from 5 to 10 pounds above normal pressure.

(3) BATTERY. Disconnect the battery to prevent its discharge by vandalism or accident. This may be accomplished by disconnecting the positive lead, taping the end of the lead, and tying it back away from the battery.

(4) BRAKES. The brakes must be applied and the transmission placed in low gear after the vehicle has been placed in position, with a brake wheel clearance of at least 6 inches (fig. 101 "A"). The vehicles will be located on the car in such a manner as to prevent the car from carrying an unbalanced load.

(5) All cars containing ordnance vehicles must be placarded "DO NOT **HUMP.**"

(6) Ordnance vehicles may be shipped on flat cars, end-door box cars, side-door box cars, or drop-end gondola cars, whichever type car is the most convenient.

b. Facilities for Loading. Whenever possible, load and unload vehicles from open cars under their own power, using permanent end ramps and spanning platforms. Movement from one flat car to another along the length of the train is made possible by cross-over plates or spanning platforms. If no permanent end ramp is available, an improvised ramp can be made from railroad ties. Vehicles may be loaded in gondola cars without drop ends by using a crane. In case of shipment in side-door cars, use a dolly-type jack to warp the vehicles into position within the car.

c. Securing Vehicles. In securing or blocking a vehicle, three motions, lengthwise, sidewise, and bouncing, must be prevented. There are two approved methods of blocking the vehicles on freight cars, as described below. When blocking dual wheels, all blocking will be located against the outside wheel of the dual.

(1) METHOD 1 (fig. 101). Locate eight blocks "B", one to the front and one to the rear of each wheel. Nail the heel of each block to the car floor, using five 40-penny nails to each block. That portion of the block under the tread will be toenailed to the car floor with two 40-penny nails to each block. Locate two blocks "D" against the outside face of each wheel. Nail the lower block to the car floor with three 40-penny nails, and the top block to the lower block with three 40-penny nails. Pass four strands, two wrappings, of No. **8** gage, black annealed wire "C" around the bumper support bracket at the front of the vehicle, and then through a stake pocket on the railroad car. Perform the same operation at the rear of the vehicle, passing the

SHIPMENT AND TEMPORARY STORAGE

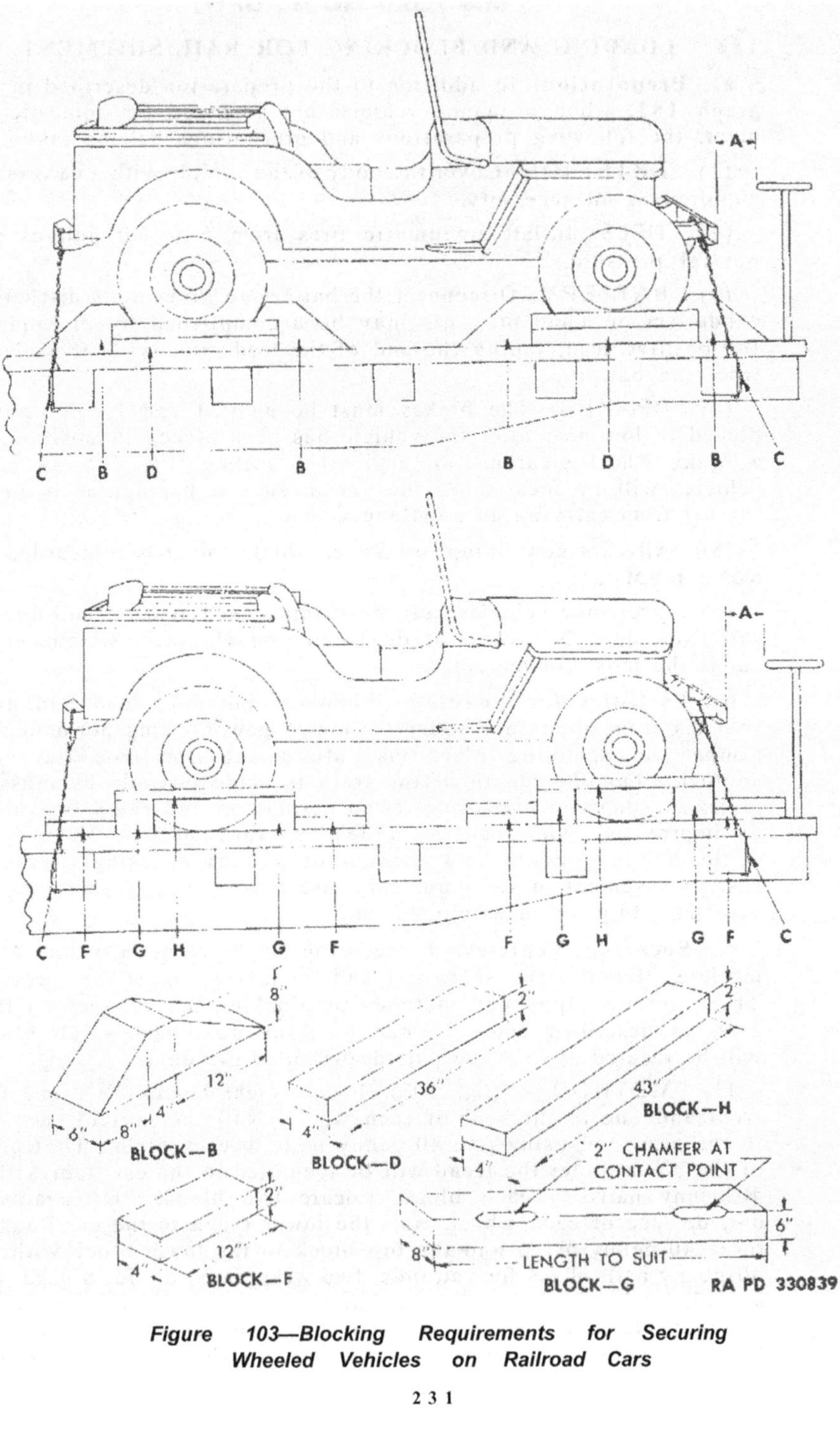

Figure 103—Blocking Requirements for Securing Wheeled Vehicles on Railroad Cars

wire through the opening in the rear bumper. Duplicate these two operations on the opposite side of the vehicle. Tighten the wires enough to remove slack. When a box car is used, this strapping must be applied in a similar fashion, and attached to the floor by the use of blocking or anchor plates. This strapping is not required when gondola cars are used.

(2) METHOD 2 (fig. 101). Place four blocks "G", one to the front and one to the rear of each set of wheels. These blocks are to be at least 8 inches wider than the over-all width of the vehicle at the car floor. Using sixteen blocks "F", locate two against blocks "G" to the front of each wheel, and two against blocks "G" to the rear of each wheel. Nail the lower cleat to the floor with three 40-penny nails, and the top cleat to the cleat below with three 40-penny nails. Locate four cleats "H" on the outside of each wheel to the top of each block "G" with two 40-penny nails. Pass four strands, two wrappings, of No. 8 gage, black annealed wire "C" around the bumper support bracket (front), and also opening in the rear bumper (rear), as described in Method 1 above.

d. **Shipping Data.**

Length, over-all	132.25 in.
Width, over-all	62 in.
Height, top down	32 in.
Shipping weight	2453 lb
Approximate floor area	57 sq ft
Approximate volume	152 cu ft
Bearing pressure (lb per sq ft)	4½

REFERENCES

PUBLICATIONS INDEXES.

The following publications indexes should be consulted frequently for latest changes to, or revisions of the publications given in this list of references and for new publications relating to materiel covered in this Manual:

Introduction to Ordnance Catalog (explains SNL system) . . . . ASF Cat. ORD-1 IOC

Ordnance publications for supply index (index to SNL's) . . . . ASF Cat. ORD-2 OPSI

Index to Ordnance publications (lists FM's, TM's, TC's, and TB's of interest to Ordnance personnel, MWO's, OPSR's, RSD, S of SR's, OSSC's, and OFSB's. Includes alphabetical listing of Ordnance major items with publications pertaining thereto) . . . . OFSB 1-1

List of publications for training (lists MR's, MTP's, T/BA*s, T/A's, FM's, TM's, and TR's, concerning training) . . . . FM 21-6

List of training films, film strips, and film bulletins (lists TF's, FS's, and FB's by serial number and subject) . . . . FM 21-7

Military training aids (lists graphic training aids, models, devices, and displays) . . . . FM 21-8

STANDARD NOMENCLATURE LISTS.

Truck 1/4-ton, 4 x 4, command reconnaissance (Ford and Willys) . . . . SNL G-503

Cleaning, preserving and lubrication materials, recoil fluids, special oils, and miscellaneous related items . . . . SNL K-1

Soldering, brazing and welding materials, gases and related items . . . . SNL K-2

Tool sets—motor transport . . . . SNL N-19

EXPLANATORY PUBLICATIONS.

Fundamental Principles.

Automotive electricity . . . . TM 10-580

Automotive lubrication . . . . TM 10-540

Basic maintenance manual . . . . TM 38-250

Driver's manual . . . . TM 10-460

Driver selection and training . . . . TM 21-300

Electrical fundamentals . . . . TM 1-455

Military motor vehicles	AR 850-15
Motor vehicle inspections and preventive maintenance service	TM 9-2810
Precautions in handling gasoline	AR 850-20
Standard Military Motor Vehicles	TM 9-2800
The internal combustion engine	TM 10-570

Maintenance and Repair.

Cleaning, preserving, lubricating and welding materials and similar items issued by the Ordnance Department	TM 9-850
Cold weather lubrication and service of combat vehicles and automotive materiel	OFSB 6-11
Maintenance and care of pneumatic tires and rubber treads	TM 31-200
Ordnance Maintenance: Engine and engine accessories for1/4-*ton* 4 x 4 truck (Ford and Willys)	TM 9-1803A
Ordnance Maintenance: Power train, chassis, and body for 1/4-ton 4 x 4 truck (Ford and Willys)	TM 9-1803B
Ordnance Maintenance: Electrical equipment (Auto-Lite)	TM 9-1825B
Ordnance Maintenance: Hydraulic brake system (Wagner)	TM 9-1827C
Ordnance Maintenance: Carburetors (Carter)	TM 9-1826A
Ordnance Maintenance: Fuel pumps	TM 9-1828A
Tune-up and adjustment	TM 10-530

Protection of Materiel.

Camouflage	FM 5-20
Chemical decontamination, materials and equipment	TM 3-220
Decontamination of armored force vehicles	FM 17-59
Defense against chemical attack	FM 21-40
Desert operations	FM 31-25
Explosives and demolitions	FM 5-25

Storage and Shipment.

Ordnance storage and shipment chart, group G—Major items	OSSC-G
Registration of motor vehicles	AR 850-10
Rules governing the loading of mechanized and motorized army equipment, also major caliber guns, for the United States Army and Navy, on open top equipment published by Operations and Maintenance Department of Association of American Railroads.	
Storage of motor vehicle equipment	AR 850-18

INDEX

B—Cont'd

INDEX

D —Cont'd

E

F

F — Cont'd

G

H

I

INDEX

R—Cont'd

S

INDEX

W

Made in the USA
Monee, IL
04 November 2026

67387219R00134